MW00843369

Digital
Multimeter
Principles

Fourth Edition

FLUKE ®

Authorized by Fluke Corporation

AMERICAN TECHNICAL PUBLISHERS
ORLAND PARK, ILLINOIS 60467-5756

Glen A. Mazur

Digital Multimeter Principles contains procedures commonly practiced in industry and the trade. Specific procedures vary with each task and must be performed by a qualified person. For maximum safety, always refer to specific manufacturer recommendations, insurance regulations, specific job site and plant procedures, applicable federal, state, and local regulations, and any authority having jurisdiction. The material contained is intended to be an educational resource for the user. Neither American Technical Publishers nor Fluke Corporation is liable for any claims, losses, or damages including property damage or personal injury incurred by reliance on this information.

Contents

Introduction

Digital Multimeters (DMMs) have become standard diagnostic tools for technicians in the electrical/electronic industry. *Digital Multimeter Principles*, 4th Edition, is designed to provide an introduction to DMM operation principles and procedures. *Digital Multimeter Principles* is authorized by the Fluke Corporation, and Fluke test tools are depicted throughout the book. Basic DMM knowledge and skills are detailed in the DMM Operation Competency Skills. Successful completion of the tasks is recognized by the Certificates of Completion.

The procedures presented in *Digital Multimeter Principles* focus on fundamental DMM operation. Additional information related to electrical/electronic testing and troubleshooting principles is available in other American Tech books. To obtain information on related training products, visit the American Tech web site at www.go2atp.com.

The Publisher

Related Training Products:

- *Electrical Motor Drive Installation and Troubleshooting*
- *Electrical Motor Controls for Integrated Systems*
- *Electrical Principles and Practices*
- *Electrical Systems Based on the 2008 NEC®*
- *Hybrid Electric Vehicle Technology*
- *Industrial Mechanics*
- *Insulation Resistance Testing*
- *Introduction to Programmable Logic Controllers*
- *Introduction to Thermography Principles*
- *Motors*
- *Photovoltaic Systems*
- *Power Quality Measurement and Troubleshooting*
- *Printreading for Installing and Troubleshooting Electrical Systems*
- *Test Instruments*
- *Transformer Principles and Applications*
- *Troubleshooting Electrical/Electronic Systems*
- *Troubleshooting Electric Motors*

DIGITALMULTIMETERPRINCIPLES

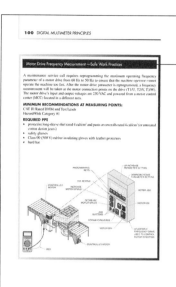

Safe work practices feature
common measurement
procedures, required PPE,
and recommended CAT ratings

Illustrated step-by-step
procedures depict
DMM procedures for
troubleshooting activities

Detailed drawings
illustrate common
troubleshooting
tasks

Safety

ELECTRICAL MEASUREMENT SAFETY

Electrical measurements are commonly taken using a digital multimeter. A *digital multimeter (DMM)* is a test tool used to measure two or more electrical values. Different DMMs can be used to measure electrical quantities such as voltage, current, resistance, and frequency. Each DMM has specific features and limits. DMMs are used in residential, commercial, and industrial electrical equipment installation and service applications.

Each application presents potential safety hazards that must be considered when taking electrical measurements. The user's manual details specific DMM specifications and features, proper operating procedures, safety precautions and warnings, and applications.

WARNING: Before using any electrical test equipment, always refer to the user's manual for proper operating procedures, safety precautions, and limits.

ELECTRICAL SHOCK

According to the National Safety Council, over 1000 people are killed by electrical shock in the United States each year.

Electricity is the number one cause of fires. Several thousand people are killed in electrical fires each year. Improper electrical wiring or misuse of electricity causes destruction of equipment and fire damage to property.

The proper safety precautions and personal protective equipment are required when using a DMM.

Electrical measurements are usually taken from exposed electric components that are normally enclosed. Safe work habits and proper equipment are required to prevent electrical shock when working with exposed electric components. *Electrical shock* is a condition that results any time a body becomes part of an electrical circuit. Electrical shock can vary in its effect from a mild sensation to fatal current. The severity of an electrical shock depends on the amount of electric current in milliamps (mA) that flows through the body, the length of time the body is exposed to the current flow, the path the current takes through the body, the amount of body area exposed to the electric contact, and the physical size and condition of the body through which the current passes. **See Figure 1-1.**

The amount of current that passes through a circuit depends on the voltage and resistance of the circuit. During an electrical shock, the body of an electrician becomes part of the circuit. The resistance that the body offers to the flow of current varies. For example, sweaty hands have less resistance than dry hands. A wet floor has less resistance than a dry floor. The lower the resistance, the greater the current flow. The greater the current flow, the greater the severity of electrical shock to the person.

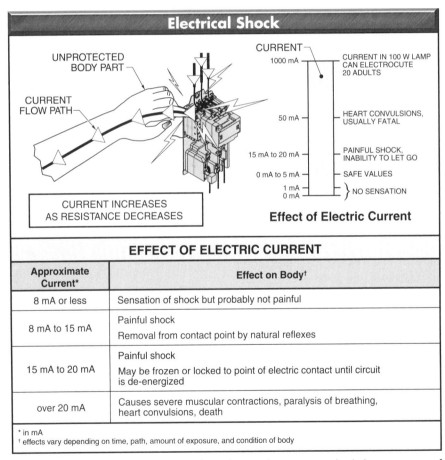

Electrical Shock

UNPROTECTED BODY PART

CURRENT FLOW PATH

CURRENT INCREASES AS RESISTANCE DECREASES

CURRENT
1000 mA — CURRENT IN 100 W LAMP CAN ELECTROCUTE 20 ADULTS
50 mA — HEART CONVULSIONS, USUALLY FATAL
15 mA to 20 mA — PAINFUL SHOCK, INABILITY TO LET GO
0 mA to 5 mA — SAFE VALUES
1 mA
0 mA — } NO SENSATION

Effect of Electric Current

EFFECT OF ELECTRIC CURRENT

Approximate Current*	Effect on Body†
8 mA or less	Sensation of shock but probably not painful
8 mA to 15 mA	Painful shock Removal from contact point by natural reflexes
15 mA to 20 mA	Painful shock May be frozen or locked to point of electric contact until circuit is de-energized
over 20 mA	Causes severe muscular contractions, paralysis of breathing, heart convulsions, death

* in mA
† effects vary depending on time, path, amount of exposure, and condition of body

Figure 1-1. Electrical shock is a condition that results any time a body becomes part of an electrical circuit.

WARNING: When taking electrical measurements, assume all electric components in a circuit are energized. Some components, such as capacitors, can cause an electrical shock from stored electrical energy even if disconnected.

When handling a victim of an electrical shock accident, apply the following procedures:

1. Break the circuit to free the victim immediately and safely. When the circuit cannot be turned OFF, use any nonconducting device to free the victim. Never touch the victim when power is not turned OFF.

2. After the victim is free from the circuit, send for help and determine if the victim is breathing. When there is no breathing or pulse, start CPR. Always get medical attention for a victim of electrical shock.

3. When the victim is breathing and has a pulse, check for burns and cuts. Burns are caused by contact with the live circuit, and are found at the points that the electricity entered and exited the body. Treat the entrance and exit burns as thermal burns and get medical help immediately.

 The National Fire Protection Association (NFPA) publishes the National Electrical Code® (NEC®). The purpose of the NEC® is the practical safeguarding of persons and property from hazards arising from the use of electricity.

PERSONAL PROTECTIVE EQUIPMENT

Personal protective equipment (PPE) is clothing and/or equipment worn by a technician to reduce the possibility of injury in the work area. The use of personal protective equipment is required whenever work may occur on or near energized exposed electrical circuits. Personal protective equipment includes protective clothing, head protection, eye protection, ear protection, hand and foot protection, back protection, knee protection, and rubber insulated matting. **See Figure 1-2.**

Clothing made of durable, fire-resistant material provides protection from contact with hot and sharp objects. Protective helmets or hard hats protect electrical workers from impact, falling and flying objects, and electrical shock. Safety glasses, respirators, ear plugs, and gloves are used based on the task performed.

For example, insulated tools and gloves made from rubber can be used to provide maximum insulation from electrical shock hazards and should be worn when taking high-voltage and current measurements. Safety shoes with steel toes provide protection from falling objects. Insulated rubber boots and rubber mats provide insulation to prevent electrical shock.

All personal protective equipment must meet Occupational Safety and Health Administration (OSHA) 29 Code of Federal Regulations (CFR) 1910, Subpart I – *Personal Protective Equipment,* sections 1910.132 – 1910.138, applicable

American National Standards Institute (ANSI) standards, and other safety standards. The National Fire Protection Association standard *NFPA 70E, Standard for Electrical Safety in the Workplace,* addresses "electrical safety requirements for employee workplaces that are necessary for the safeguarding of employees in pursuit of gainful employment."

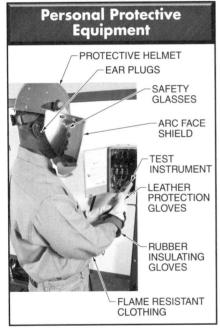

Personal Protective Equipment

PROTECTIVE HELMET
EAR PLUGS
SAFETY GLASSES
ARC FACE SHIELD
TEST INSTRUMENT
LEATHER PROTECTION GLOVES
RUBBER INSULATING GLOVES
FLAME RESISTANT CLOTHING

Figure 1-2. Personal protective equipment (PPE) is used when taking electrical measurements to reduce the possibility of an injury.

NFPA 70E

For maximum safety, personal protective equipment and safety requirements for DMM procedures must be followed as specified in NFPA 70E, OSHA Standard Part 1910 *Subpart I—Personal Protective Equipment* (1910.132 through 1910.138), and other applicable safety mandates. Per NFPA 70E, "Only qualified persons shall perform testing on or near live parts operating at 50 V or more."

All personal protective equipment and tools are selected to be appropriate to the operating voltage (or higher) of the equipment or circuits being worked on or near. Equipment, devices, tools, and DMMs must be suited for the work to be performed. In most cases, voltage-rated gloves and tools are required. Voltage-rated gloves and tools are rated and tested for the maximum line-to-line voltage upon which work will be performed. Protective gloves must be inspected or tested as required for maximum safety before each task.

Rubber Insulating Gloves

Rubber insulating gloves are an important article of personal protective equipment for electrical workers. Safety requirements for the usage of rubber insulating gloves and cover gloves (commonly leather) must be followed at all times. The primary purpose of rubber insulating gloves and cover gloves is to insulate hands and lower arms from possible contact with live conductors.

Rubber insulating gloves offer a high resistance to current flow to help prevent an electrical shock and the leather protectors protect the rubber insulating glove and provide additional insulation. Rubber insulating gloves are rated and labeled for maximum voltage allowed.

WARNING: Rubber insulating gloves are designed for specific applications. Leather protectors are required for protecting rubber insulating gloves. Rubber insulating gloves offer the highest resistance and greatest insulation. Serious injury or death can result from improper use of rubber insulating gloves, or from using outdated and/or the wrong type of rubber insulating gloves for an application.

Any substance that can physically damage rubber gloves must be removed before testing. Insulating gloves and protector gloves found to be defective shall not be discarded or destroyed in the field, but shall be tagged "unsafe" and returned to a supervisor.

Flame-Resistant (FR) Clothing

Sparks from an electrical circuit can cause a fire. Approved flame-resistant (FR) clothing must be worn in conjunction with rubber insulating gloves for protection from electrical arcs when performing certain operations on or near energized equipment or circuits. FR clothing must be kept as clean and sanitary as practical and must be inspected prior to each use. Defective clothing must be removed from service immediately and replaced. Defective FR clothing must be tagged "unsafe" and returned to a supervisor.

Eye Protection

Eye protection must be worn to prevent eye or face injuries caused by flying particles, contact arcing, and radiant energy. Eye protection must comply with OSHA 29 CFR 1910.133, *Eye and Face Protection.* Eye protection standards are specified in ANSI Z87.1, *Occupational and Educational Eye and Face Protection.* Eye protection includes safety glasses, goggles, face shields, and arc blast hoods. **See Figure 1-3.**

Safety glasses are an eye protection device with special impact-resistant glass or plastic lenses, reinforced frames, and side shields. Plastic frames are designed to keep the lenses secured in the frame when an impact occurs to minimize the shock hazard when working with electrical equipment. Side shields provide additional protection from flying objects. Tinted-lens safety glasses protect against low-voltage arc hazards. Non-tinted safety glasses with side shields can be worn when working on low-voltage (50 V or less) PLC systems that are operating in a normally safe environment.

Rubber insulating gloves and leather protectors provide protection from electrical arcs when working near energized equipment or circuits.

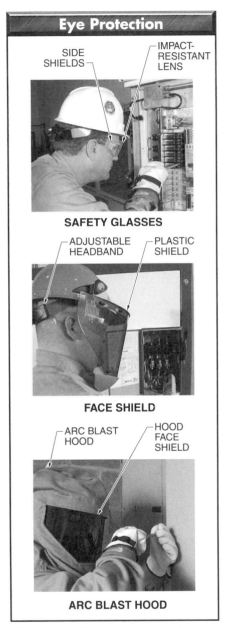

Eye Protection

SIDE SHIELDS

IMPACT-RESISTANT LENS

SAFETY GLASSES

ADJUSTABLE HEADBAND

PLASTIC SHIELD

FACE SHIELD

ARC BLAST HOOD

HOOD FACE SHIELD

ARC BLAST HOOD

Figure 1-3. Eye protection must be worn to prevent eye or face injuries caused by flying particles, contact arcing, or radiant energy.

Goggles are an eye protection device with a flexible frame that is secured on the face with an elastic headband. Goggles fit snugly against the face to seal the areas around the eyes, and can be used over prescription glasses. Goggles with clear lenses protect against small flying particles or splashing liquids. Tinted goggles are sometimes used to protect against low-voltage arc hazards.

A *face shield* is any eye and face protection device that covers the entire face with a plastic shield, and is used for protection from flying objects. Tinted face shields protect against low-voltage arc hazards.

An *arc blast hood* is an eye and face protection device that covers the entire head with plastic and material. Arc blast hoods are used to protect against high-voltage arc blasts. Technicians working with energized high-voltage equipment must wear arc blast protection.

Safety glasses, goggles, face shields, and arc blast hood lenses must be properly maintained to provide protection and clear visibility. Lens cleaners are available that clean without risk of lens damage. Pitted or scratched lenses reduce vision and may cause lenses to fail on impact.

Rubber Insulating Matting

Rubber insulating matting is a personal protective device placed on the floor to provide electricians protection from electrical shock when working on energized electrical circuits. Dielectric black fluted rubber matting is specifically designed for use in front of open cabinets or high-voltage equipment. Rubber insulating matting is used to protect electricians when voltages are over 50 V. Rubber insulating matting is available in different sizes and is rated for maximum working voltage. **See Figure 1-4.**

INTERNATIONAL ELECTRO-TECHNICAL COMMISSION (IEC) 61010 SAFETY STANDARD

The *International Electrotechnical Commission (IEC)* is an organization that develops international safety standards for electrical equipment. The IEC standards reduce safety hazards that can occur from unpredictable circumstances when using electrical test equipment such as DMMs. For example, voltage surges on a power distribution system can cause a safety hazard when a DMM or other test instruments are being used in an electrical system.

A *voltage surge* is a higher-than-normal voltage that temporarily exists on one or more power lines. Voltage surges vary in voltage amount and time present on the power lines. One type of voltage surge is a transient voltage. A *transient voltage* (voltage spike) is a temporary, undesirable voltage in an electrical circuit. Transient voltages typically exist for a very short time, but are large in magnitude and very erratic. Transient voltages occur due to lightning strikes, unfiltered electrical equipment, and power being switched ON and OFF.

A surge suppressor is an electrical device that can provide protection from high-level transients by limiting the level of voltage allowed downstream.

Rubber Insulating Matting Ratings					
Safety Standard	Material Thickness		Material Width (in.)	Test Voltage	Maximum Working Voltage
	Inches	Millimeters			
BS921*	0.236	6	36	11,000	450
BS921*	0.236	6	48	11,000	450
BS921*	0.354	9	36	15,000	650
BS921*	0.354	9	48	15,000	650
VDE0680[†]	0.118	3	39	10,000	1000
ASTM D178[‡]	0.236	6	24	25,000	17,000
ASTM D178[‡]	0.236	6	30	25,000	17,000
ASTM D178[‡]	0.236	6	36	25,000	17,000
ASTM D178[‡]	0.236	6	48	25,000	17,000

* BSI–British Standards Institute
[†] VDE–Verband Deutscher Elektrotechniker Testing and Certification Institute
[‡] ASTM International

Figure 1-4. Rubber insulating matting provides protection from electrical shock when working on energized electrical circuits.

To help protect personnel, DMMs, and other test instruments from dangerous voltages, the IEC sets standards for distances between mounted components on a board (creepage distances) and distances between components through the air (clearance distances) inside a meter. The higher the working voltage, the greater the required distances.

High transient voltages can reach several thousand volts. A transient voltage on a 120 V power line can reach 1000 V (1 kV) or more. High transient voltages exist close to a lightning strike or when large (high-current) loads are switched OFF. **See Figure 1-5.** For example, when a large motor (100 HP) is turned OFF, a transient voltage moves down the power distribution system. When a DMM is connected to a point along the system in which a high transient voltage is present, an arc can be created inside the meter. Once started, the arc can cause a high-current short in the power distribution system even after the original high transient voltage is gone. A high-current short can turn into an arc blast.

An *arc blast* is an explosion that occurs when the air surrounding electrical equipment becomes ionized and conductive. The amount of current drawn and the potential damage caused depend on the specific location of the arc blast in the power distribution system. All power distribution systems have current limits set by fuses and circuit breakers along the system. The current rating (size) of the fuses and circuit breakers decreases further away from the main distribution panel. The further away from the main distribution panel, the less likely a high transient voltage is to cause damage.

CAUTION: Other than a laser, an electric arc is the hottest heat source in existence. Electric arcs are capable of producing temperatures up to 10,000°F. Temperatures of such intensity are capable of producing serious burns at distances up to 20′ and can be fatal at distances up to 8′.

Measurement Categories

IEC standard 61010 classifies the applications in which DMMs may be used into four measurement categories (Category I – Category IV). The four categories are typically abbreviated as CAT I, CAT II, CAT III, and CAT IV. The CAT ratings determine what magnitude of transient voltage a DMM or other electrical appliance can withstand when used on a power distribution system.

For example, a DMM specified for use in a CAT III installation must withstand a 6000 V transient (2 ms rise time and a 50 ms, 50% duration) voltage without resulting in a hazard. When a DMM is operated on voltages above 600 V, the DMM must be capable of withstanding an 8000 V transient voltage. The meters must withstand a total of 20 transient voltages—10 positive transients and 10 negative transients for the respective transient level. Also, a DMM that can withstand these transient voltages can be damaged but the transient cannot result in a hazard to the technician or the facility. To protect technicians from transient voltages, protection must be built into all test equipment.

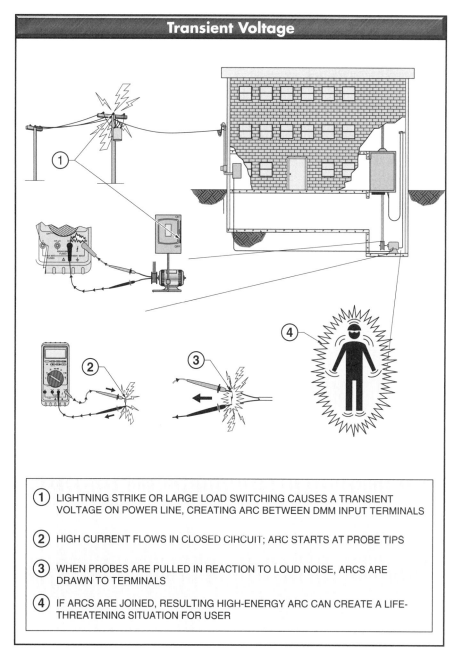

Transient Voltage

1. LIGHTNING STRIKE OR LARGE LOAD SWITCHING CAUSES A TRANSIENT VOLTAGE ON POWER LINE, CREATING ARC BETWEEN DMM INPUT TERMINALS

2. HIGH CURRENT FLOWS IN CLOSED CIRCUIT; ARC STARTS AT PROBE TIPS

3. WHEN PROBES ARE PULLED IN REACTION TO LOUD NOISE, ARCS ARE DRAWN TO TERMINALS

4. IF ARCS ARE JOINED, RESULTING HIGH-ENERGY ARC CAN CREATE A LIFE-THREATENING SITUATION FOR USER

Figure 1-5. When taking measurements in an electrical circuit, transient voltages can cause electrical shock and/or damage to equipment.

Safety standards such as IEC 61010-1 2nd edition, the harmonized North America standard, and the UL standard 61010B vary, but are closely matched. The requirements of the standards are used to rate test equipment for minimizing hazards such as shock, fire, and arc blast among other concerns. A DMM that meets these standards offers a high level of protection.

A measurement category rating such as CAT III or CAT IV indicates acceptable usage on three-phase permanently installed loads and three-phase distribution panels in a building or facility. All exposed electrical installations and the power panels of a facility are considered high-voltage areas. Measurement categories such as CAT III and CAT IV ratings are important criteria for DMMs used in industrial applications. A higher CAT number indicates an electrical environment with higher power available, larger short-circuit current available, and higher energy transients. For example, a DMM that meets the CAT III standard is resistant to higher energy transients than a DMM that meets the CAT II standard. **See Figure 1-6.**

Power distribution systems are divided into categories based on the magnitude of transient voltage DMMs must withstand when used on the power distribution system. Dangerous high-energy transient voltages such as a lightning strike are attenuated (lessened) or dampened as the transient travels through the impedance (AC resistance) of the system and system grounds

Within an IEC 61010 standard category, a higher voltage rating denotes a higher transient voltage withstanding rating. For example, a CAT III-1000 V (steady-state)

rated DMM has better protection compared to a CAT III-600 V (steady-state) rated DMM. But, a CAT III-600 V DMM has equivalent withstanding capability to a CAT II 1000 V DMM. A DMM must be chosen based on the IEC measurement category first and voltage second. *Note: IEC standards for test instruments are not the same as the hazard categories defined by NFPA 70E.*

INDEPENDENT TESTING ORGANIZATIONS

National, state, and local codes and standards are used to protect people and property from electrical dangers. A *code* is a regulation or minimum requirement. A *standard* is an accepted reference or practice. Codes and standards ensure that electrical equipment is built and installed safely and that every effort is made to protect people from electrical shock. The IEC sets standards but does not test or inspect for code and standard compliance.

Always use proper PPE and the correct CAT rated DMM for the measurement application.

IEC 61010 Measurement Categories

Category	In Brief	Examples
CAT I	Electronic	• Protected electronic equipment • Equipment connected to (source) circuits in which measures are taken to limit transient overvoltages to an appropriately low level • Any high-voltage, low-energy source derived from a high-winding resistance transformer, such as the high-voltage section of a copier
CAT II	1φ receptacle-connected loads	• Appliances, portable tools, and other household and similar loads • Outlets and long branch circuits • Outlets at more than 30′ (10 m) from CAT III source • Outlets at more than 60′ (20 m) from CAT IV source
CAT III	3φ distribution, including 1φ commercial lighting	• Equipment in fixed installations, such as switchgear and polyphase motors • Bus and feeder in industrial plants • Feeders and short branch circuits and distribution panel devices • Lighting systems in larger buildings • Appliance outlets with short connections to service entrance
CAT IV	3φ at utility connection, any outdoors conductors	• Refers to the origin of installation, where low-voltage connection is made to utility power • Electric meters, primary overcurrent protection equipment • Outside and service entrance, service drop from pole to building, run between meter and panel • Overhead line to detached building

Figure 1-6. The IEC 61010 standard classifies the application in which a DMM may be used into four measurement categories.

A DMM with a symbol and listing number of an independent testing lab such as Underwriters Laboratories Inc.® (UL), Canadian Standards Association (CSA), or other recognized testing organization indicates compliance with the standards of the organization. **See Figure 1-7.** A manufacturer can claim to "design to" a standard with no independent verification. To be UL listed or CSA certified, a manufacturer must employ the services of an approved agency to test a product's compliance with a standard. **See Figure 1-8.** For example, UL 61010B or CSA C22.2 No. 1010.1-92 indicates that the IEC 61010 standard is met.

WARNING: Before using any DMM, always refer to the user's manual for proper operating procedures, safety precautions, and technical limits. Conditions can change quickly as voltage and current levels vary in individual circuits.

National Fire Protection Association (NFPA)

The *NFPA* is a national organization that provides guidance in assessing the hazards of the products of combustion. The NFPA publishes the National Electrical Code® (NEC®). The purpose of the NEC® is the practical safeguarding of persons and property from the hazards arising from the use of electricity. The NEC® is updated every three years. Many city, county, state, and federal agencies use the NEC® to set requirements for electrical installations. The primary concern of the NEC® is to protect lives and property.

Occupational Safety and Health Administration (OSHA)

The *Occupational Safety and Health Administration (OSHA)* is a federal government agency established under the Occupational Safety and Health Act of 1970, which requires all employers to provide a safe environment for their employees. The Act requires that all employers provide work areas free from recognized hazards likely to cause serious harm.

The primary concern of NFPA 70E is to meet OSHA safety standards and comply with NEC® requirements. OSHA mandates that there shall be a safe work environment, and NFPA 70E and the NEC® states how it shall be done.

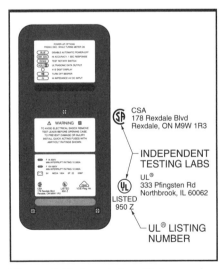

Figure 1-7. A symbol and listing number indicates compliance with IEC standards.

A DMM with a clamp-on current probe accessory is used to measure current.

Recognized Testing Laboratories (RTLs) and Standards Organizations*

Underwriters Laboratories, Inc.® (UL)	2600 N.W. Lake Rd. Camas, WA 98607-8542 Tel: 877-854-3577 www.ul.com
American National Standards Institute (ANSI)	25 W. 43rd St., 4th Floor New York, NY 10036 Tel: 212-642-4900 www.ansi.org
British Standards Institution (BSI)	389 Chiswick High Road London W4 4AL United Kingdom www.bsigroup.com
CENELEC Comité European de Normalisation Electrotechnique	17 Avenue Marnix B-1000 Brussels, Belgium Tel: 416-747-4000 www.cenelec.eu
Canadian Standards Association (CSA)	Central Office 178 Rexdale Blvd. Etobicoke (Toronto), Ont. M9W 1R3 Tel: 416-747-4000 www.csa.ca
Verband der Elektrotechnik und Informationstechnik (VDE)	Frankfurt am main Germany www.vde.de
Japanese Standards Association (JSA)	1-24, Akasaka 4 Minato-ku Tokyo 107 Japan
International Electrotechnical Commission (IEC)	3, rue de Varembé PO Box 131 CH-1211 Geneva 20 Switzerland Tel: +41 22 919 02 11 www.iec.ch
The Institute of Electrical and Electronic Engineers, Inc. (IEEE)	3 Park Avenue,17th Floor New York, NY 10016-5997 Tel: 800-678-4333 www.ieee.org
National Institute of Standards and Technology Calibration Program (NIST)	100 Bureau Drive, Stop 2300 Gaithersburg, MD 20899-2300 Tel: 301-975-2200 www.nist.gov
National Electrical Manufacturers Association (NEMA) Standards Publication Office	1300 North 17th Street, Suite 1752 Rosslyn, Virginia 22209 Tel: 703-841-3200 www.nema.org
International Standards Organization (ISO)	1 rue de Varembé Case postale 56 CH-1211 Geneva 20 Switzerland Tel: +41 22 749 01 11 www.iso.ch
OSHA Region 1 Regional Office	JFK Federal Building, Room E340 Boston, MA 02203 Tel: 617-565-9860 www.osha.gov
TÜV Rheinland of North America, Inc.	12 Commerce Rd., Newton, CT 06470 Tel: 203-426-0888 www.us.tuv.com
Australian Communications and Media Authority (ACMA)	Level 15, Tower 1 Darling Park 201 Sussex Street, Sydney NSW 2000 Tel: 02 9334 7700 www.acma.gov.au

* partial listing

Figure 1-8. To be UL listed or CSA certified, a manufacturer must employ the services of an approved agency to test product compliance with the specific standard.

National Electrical Manufacturers Association (NEMA)

NEMA is a national organization that assists with information and standards concerning proper selection, ratings, construction, testing, and performance of electrical equipment. NEMA standards are used as guidelines for the manufacture and use of electrical equipment.

GROUNDING

Electrical circuits are grounded to safeguard equipment and personnel against the hazards of electrical shock. Proper grounding of electrical tools, machines, equipment, and delivery systems is one of the most important factors in preventing hazardous conditions. *Grounding* is the connection of all exposed non-current-carrying metal parts to the earth. Grounding provides a direct path for unwanted (fault) current to the earth without causing harm to persons or equipment.

Non-current-carrying metal parts that are connected to ground include all metal boxes, raceways, enclosures, and equipment. Unwanted current exists because of insulation failure or because a current-carrying conductor makes contact with a non-current-carrying part of the system. In a properly grounded system, the unwanted current flow trips fuses or circuit breakers. Once the fuse or circuit breaker is tripped, the circuit is opened and no additional current flows.

LOCKOUT/TAGOUT

Lockout is the process of removing the source of electrical power and installing a lock that prevents the power from being turned ON. To ensure the safety of personnel working with equipment, all electrical, pneumatic, and hydraulic power is removed and the equipment must be locked out and tagged out. *Tagout* is the process of placing a danger tag on the source of electrical power, which indicates that the equipment may not be operated until the danger tag is removed. Per OSHA standards, equipment is locked out and tagged out before any installation or preventive maintenance is performed. **See Figure 1-9.**

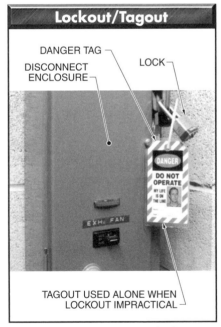

Lockout/Tagout

DANGER TAG

DISCONNECT ENCLOSURE

LOCK

TAGOUT USED ALONE WHEN LOCKOUT IMPRACTICAL

Figure 1-9. Per OSHA standards, equipment must be locked out and tagged out before any installation or preventive maintenance is performed.

A danger tag has the same importance and purpose as a lock and is used alone only when a lock does not fit the disconnect device. A danger tag must be attached at the disconnect device with a tag tie or equivalent and must have space for the technician's name, craft, and other company-required information. A danger tag must withstand the elements and expected atmosphere for the maximum period of time that exposure is expected.

Lockout/tagout is used in the following situations:

* power is not required to be ON for a piece of equipment to perform a task

* when removing or bypassing machine guards or other safety devices

* the possibility exists of being injured or caught in moving machinery

* when jammed equipment is being cleared

* the danger exists of being injured if equipment power is turned ON

ARC BLAST SAFETY

An *electric arc* is a discharge of electric current across an air gap. Arcs are caused by excessive voltage ionizing an air gap between two conductors, or by accidental contact and reseparation between two conductors. When an electric arc occurs, there is the possibility of "arc flash" or "arc blast."

An *arc flash* is an electrical discharge that occurs when electrical current passes through the air separating energized/hot (ungrounded) conductors and ground (grounded parts). Arc flashes produce extremely high temperatures and can set clothes on fire, cause severe burns, and can result in death.

An *arc blast* is an explosion that occurs when the air surrounding electrical equipment becomes ionized and conductive. Arc blasts are a threat to electrical systems of 480 V and higher. Arc blasts are possible in systems of lesser voltage, but arc blasts are not likely to be as destructive as in a high-voltage system.

Arc flash and arc blast are always a possibility when working with DMMs or other test instruments. Arc flash can occur when using a DMM to measure voltage in a 480 V or higher electrical system when there happens to be a power line transient, such as a lightning strike or power surge, at the same time. Some DMMs indicate a CAT rating specifying the tolerance limit for overvoltage transients and subsequent safety provisions. CAT rated DMMs are designed to minimize or reduce the possibility of an arc flash occurring inside of the DMM.

A potential cause for arc flash and arc blast is improper DMM use. For example, an arc blast can occur when connecting a DMM across two points of a circuit that is energized with a higher voltage than the rating of the meter. To avoid causing arc blast or arc flash, an electrical system needs to be de-energized, locked out, and tagged out prior to performing work. Only qualified electricians are allowed to work on energized circuits of 50 V or higher.

Flash Protection Boundary

The *flash protection boundary* is the distance at which PPE is required to prevent burns when an arc occurs. **See Figure 1-10.** Per NFPA 70E, systems of 600 V and less require a flash protection boundary of 4′, based on the time of the circuit breaker to act. While a circuit that is being worked on should always be de-energized, the possibility exists that nearby circuits are still energized and within the flash point boundary. Barriers such as insulation blankets, along with the proper PPE, must be used to protect against flashing from nearby energized circuits.

NFPA 70E sets flash protection boundaries, in which required PPE must be worn, and states that all test equipment and accessories used within the boundary must be rated for the specific electrical environment. For example, a DMM rated for CAT III 600 V and proper PPE are required when taking live measurements in an NFPA 70E Hazard/Risk Category 2 (240 V to 600 V) electrical environment.

Personal Protective Equipment (PPE) for Arc Blast Protection. Proper PPE must always be worn when working with energized electrical circuits. Clothing made of synthetic materials such as nylon, polyester, or rayon, alone or combined with cotton, must never be worn as synthetic materials burn and melt to the skin. Per NFPA 70E, the type of PPE worn depends on the type of work being performed. The minimum PPE requirement for electrical work is an untreated natural material long-sleeve shirt, long pants, safety glasses with side shields, and rubber insulating shoes or boots.

Additional PPE includes flame-retardant (FR) coveralls, FR long-sleeve shirts and pants, a hard hat with an FR liner, hearing protection, and double-layer flash suit jacket and pants. Flash suits are similar to firefighter turnout gear and must be worn when working near a Category IV Hazard/Risk area. **See Figure 1-11.**

 Electricity is the number one cause of fires in the United States.

Approach Boundaries to Energized Parts for Shock Prevention

Nominal System Voltage, Range, Phase to Phase*	Limited Approach Boundary		Restricted Approach Boundary (Allowing for Accidental Movement)	Prohibited Approach Boundary
	Exposed Movable Conductor	Exposed Fixed-Circuit Part		
0 to 50	N/A	N/A	N/A	N/A
51 to 300	10′-0″	4′-0″	Avoid Contact	Avoid Contact
301 to 750	10′-0″	4′-0″	1′-0″	0′-1″
751 to 15,000	10′-0″	4′-0″	2′-2″	0′-7″

* in volts

Figure 1-10. The approach boundary is the distance at which PPE is required while working on energized circuits to prevent burns if an arc occurs.

Flame-Resistant Protective Equipment Requirements

Flame-Retardant Clothing Type	NFPA 70E Category Number (1 = Least Hazardous)				
	1	2	2*	3	4
Required minimum arc rating of PPE (CAL/CM²)	4	8	8	25	40
Flash suit jacket					X
Flash suit pants					X
Head protection (hard hat)	X	X	X	X	X
Flame-retardant hard hat liner				X	X
Safety glasses w/side shields or goggles	X	X	X	X	X
Arc faceshield with wrap-around face, forehead, ear, and neck protection		X	X	X	X
Face protection (2-layer hood)			X	X	X
Hearing protection (ear canal inserts)			X	X	X
Rubber gloves w/leather protectors	X	X	X	X	X
Leather shoes w/rubber soles	X	X	X	X	X

* is a higher energy environment than Category 2

Figure 1-11. Per NFPA 70E, the type of PPE required depends on the voltage and where work is being performed.

DMM SAFETY PRECAUTIONS

Conditions can change quickly as voltage and current levels vary in individual circuits. General safety precautions required when using DMMs include the following:

- When a circuit does not have to be energized (as when taking a resistance measurement or checking diodes and capacitors in a circuit), lockout and tagout all equipment and circuits to be tested.
- Never assume a DMM is operating correctly. Check the DMM that will be measuring voltage on a known (energized) voltage source before taking a measurement on an unknown voltage source. After taking a measurement on an unknown voltage source, retest the DMM on a known source to verify the meter still operates properly. This prevents a malfunctioning DMM or blown fuse in the DMM from giving a false (no

voltage) reading on an energized circuit. This check is known as the three-point meter safety test. This test verifies the integrity of the meter and meter test leads and is required by NFPA 70E.

- Always assume that equipment and circuits are energized until positively identified as de-energized by taking proper measurements.
- Never work alone when working on or near exposed energized.
- Always wear personal protective equipment (safety glasses, rubber insulating gloves, cover gloves, arc flash protection) appropriate for the procedure performed and work area. **See Figure 1-12.**
- Ensure that the function switch is set to the proper range and function before applying test leads to a circuit. A DMM set to the wrong function can be damaged. For example, possible damage can occur if test leads contact

an AC power source while the meter is set to measure resistance or continuity.

Procedural Safety
⚠ WARNING

- Follow all electrical safety practices and procedures
- Check and wear personal protective equipment (PPE) for the procedure being performed
- Perform only authorized procedures
- Follow all manufacturer recommendations and procedures

Figure 1-12. The four warning procedures used in procedural lists are typically used to encompass many aspects of safety for an electrician.

- Ensure that the test leads are connected properly. Test leads that are not connected to the correct jacks can be dangerous. For example, attempting to measure voltage while the test leads are in the amps jack produces a short circuit. For maximum safety, use a DMM that is self-protected with a high-energy fuse. Follow the operating instructions in the user's manual or the directions on the graphic display for proper test lead connections and operating information.
- Start with the highest range when measuring unknown values. Using a range too low can damage the DMM. Likewise, attempting a voltage or current measurement above the rated limit can be dangerous.
- Connect the ground test lead (black) first, the voltage test lead (red) next. Disconnect the voltage test lead (red) first, and the ground test lead (black) next after taking measurements.

- Whenever possible, connect test leads to the output side (load side) of a circuit breaker or fuse to provide better short circuit protection.
- Never assume that a circuit is de-energized or equipment is fully discharged. Capacitors can hold a charge for a long time—several minutes or more. Always check for the presence of voltage before taking any other measurements.
- Check test leads for frayed or broken insulation. Test leads are considered wear items as they will wear out with use, even with proper care. Consideration should be given to replacing test leads regularly if they are subjected to heavy use. Doing so will further protect the user from false readings due to heavily worn test leads. Always check for continuity of the test leads before use. Electrical shock can occur from accidental contact with live components. Electrical test equipment should have double-insulated test leads, recessed input jacks on the meter, shrouds on the test lead plugs, and finger guards on test probes.
- Use DMMs that conform to the IEC 61010 category in which they will be used. For example, to measure 480 V in an electrical distribution feeder panel, a DMM rated at CAT III-600 V, 1000 V or CAT IV-600 V is used.
- Avoid taking measurements in humid or damp environments.
- Ensure that no atmospheric hazards such as flammable dust or vapor is present in the area.
- Use one hand when working on a live circuit to reduce the chance of an electrical shock passing through the heart and lungs.

DMM Abbreviations, Symbols, and Terminology

DMM TYPES

DMMs are available with a wide assortment of functions and features ranging from a few measuring functions to very specialized measuring and recording features. DMMs can be broadly classified as general-purpose, standard, or advanced DMMs. **See Figure 2-1.**

General-Purpose DMMs (Testers)

A general-purpose DMM (tester) is typically used for taking basic voltage, resistance, and current (some models) measurements when testing basic components (switches, fuses, etc.) or taking basic troubleshooting measurements (voltage level, etc.). Characteristics of a general-purpose DMM include the following:

- ability to measure two or more basic electrical quantities (voltage, resistance, and current)

- a continuity test mode, excluding any specialized measuring features such as testing diodes or features to measure capacitance, temperature, or frequency

- a display usually limited to three digits

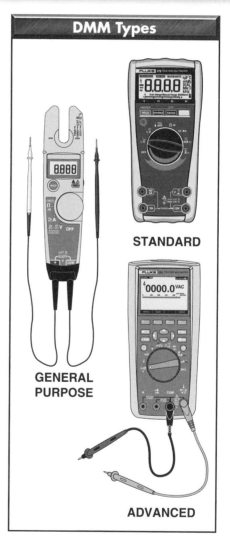

DMM Types

STANDARD

GENERAL PURPOSE

ADVANCED

Figure 2-1. DMMs are typically classified as general-purpose, standard, or advanced.

Standard DMMs

A standard DMM is useful for taking measurements and troubleshooting because it includes many features for different applications such as residential, commercial, and industrial applications; automotive applications; aviation and marine applications; and medical equipment applications. Characteristics of a standard DMM include the following:

- ability to measure three or more basic electrical quantities (voltage, resistance, and current)

- several specialized test modes such as testing diodes or features to measure capacitance, temperature, or frequency

- helpful features such as a backlight display for dark areas and input alert to warn when test lead connections do not match selector switch setting (e.g., test lead in AMP jacks and selector switch set to Voltage)

- usage of add-on attachments such as specialized test leads, current clamps, temperature attachments, and pressure attachments

- extra selector switch positions, such as Low-Impedance setting, mV (DC and/or AC), and mA/µA (DC and/or AC), to expand and improve measurement accuracy

- recording/memory functions such as MIN MAX, Average, Peak, and Relative Recording modes; recorded measurements can be retrieved on the display but cannot be downloaded to a PC for storage, documentation, or printing

- a display usually limited to four digits

Advanced DMMs

An advanced DMM can be used for standard DMM applications, troubleshooting complex problems, and identifying conditions that can lead to problems (predictive maintenance). It can also be used when documentation and printed data are required. Characteristics of an advanced DMM include the following:

- standard DMM features

- four-digit display with higher count capability or five-digit display

- specialized measuring features such as those that directly measure crest factor and low resistance for troubleshooting power quality problems, motor windings, or electrical contacts

- screen data that can be saved in a file within the meter and viewed at a later time

- may include a real time clock for time stamping of recorded data

- a trending feature that allows measurements taken over time to be displayed as a single line on a graph

- connection capability to a PC so that recorded data can be downloaded, documented, and printed

DMM USAGE

A DMM is commonly used for measuring electrical quantities such as voltage, amperage, or resistance in electrical and electronic circuits. The DMM user's manual details specific DMM specifications, features, operating procedures, safety precautions, and applications. Although DMM features and operations may vary, most abbreviations, symbols, and terminology are standardized in the industry. Abbreviations and symbols are used to simplify the expression of common electrical and electronic quantities.

DMM Abbreviations

An *abbreviation* is a letter or combination of letters that represents a word. Abbreviations used are dependent on a particular language. Generally, abbreviations which spell a word are followed by a period. DMMs commonly use standard abbreviations to represent a quantity or term for quick recognition. For example, quantities such as voltage and current are identified with abbreviations. Abbreviations can be

used individually or in combination with prefixes such as millivolt (mV) or milliamp (mA). **See Figure 2-2.**

DMM Symbols

A *symbol* is a graphic element that represents a quantity, unit, or component. Symbols provide quick recognition and are independent of language because a symbol can be interpreted regardless of the language a person speaks. For example, standard symbols commonly used on DMMs can represent an electrical component (diode, battery, etc.), term (AC, positive, etc.), or message to the user (warning, etc.). Some DMM functions are represented by two symbols combined. For example, a diode test is represented by the symbols for diode and audio beeper. **See Figure 2-3.**

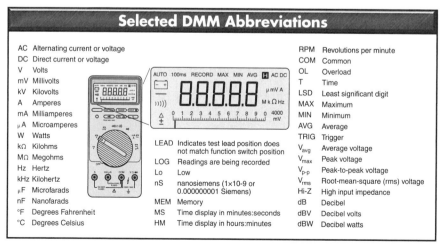

Selected DMM Abbreviations

AC	Alternating current or voltage		RPM	Revolutions per minute
DC	Direct current or voltage		COM	Common
V	Volts		OL	Overload
mV	Millivolts		T	Time
kV	Kilovolts		LSD	Least significant digit
A	Amperes		MAX	Maximum
mA	Milliamperes		MIN	Minimum
μA	Microamperes		AVG	Average
W	Watts		TRIG	Trigger
kΩ	Kilohms	LEAD	Indicates test lead position does not match function switch position	V_{avg}	Average voltage
MΩ	Megohms		V_{max}	Peak voltage
Hz	Hertz	LOG	Readings are being recorded	V_{p-p}	Peak-to-peak voltage
kHz	Kilohertz	Lo	Low	V_{rms}	Root-mean-square (rms) voltage
μF	Microfarads	nS	nanosiemens (1×10-9 or 0.000000001 Siemens)	Hi-Z	High input impedance
nF	Nanofarads	MEM	Memory	dB	Decibel
°F	Degrees Fahrenheit	MS	Time display in minutes:seconds	dBV	Decibel volts
°C	Degrees Celsius	HM	Time display in hours:minutes	dBW	Decibel watts

Figure 2-2. Abbreviations are letters or combinations of letters used to represent a word.

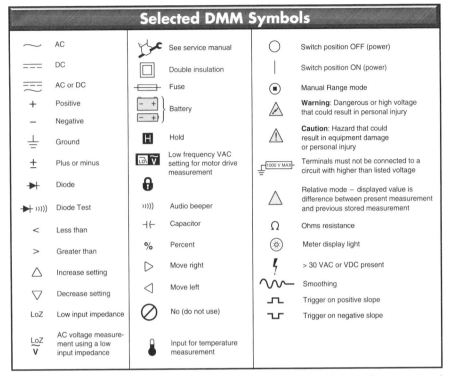

Selected DMM Symbols

Symbol	Meaning	Symbol	Meaning	Symbol	Meaning
~	AC		See service manual	○	Switch position OFF (power)
---	DC	□	Double insulation	|	Switch position ON (power)
~=	AC or DC		Fuse	⊙	Manual Range mode
+	Positive		Battery	⚠	**Warning**: Dangerous or high voltage that could result in personal injury
−	Negative				**Caution**: Hazard that could
⏚	Ground	**H**	Hold	⚠	result in equipment damage or personal injury
±	Plus or minus		Low frequency VAC setting for motor drive measurement	1000 V MAX	Terminals must not be connected to a circuit with higher than listed voltage
▸⊢	Diode	🔒		△	Relative mode − displayed value is difference between present measurement and previous stored measurement
▸⊢)))	Diode Test	)))	Audio beeper		
<	Less than	⊣⊢	Capacitor	Ω	Ohms resistance
>	Greater than	%	Percent	⊙	Meter display light
△	Increase setting	▷	Move right	⚡	> 30 VAC or VDC present
▽	Decrease setting	◁	Move left	∿	Smoothing
LoZ	Low input impedance	⊘	No (do not use)	⊓	Trigger on positive slope
LoZ/V	AC voltage measurement using a low input impedance	🌡	Input for temperature measurement	⊔	Trigger on negative slope

Figure 2-3. Symbols represent a quantity, unit, or component that can be recognized regardless of the language a person speaks.

DMM Terminology

All DMMs use specific terms to describe displayed information. These terms may be abbreviated or shown as a symbol. For example, a DMM may display an electrical quantity as an average, effective, peak, peak-to-peak, or rms value. The rms value is the root-mean-square value. It is always equal to 0.707 of a pure sine waveform amplitude. **See Figure 2-4.**

The theory of induction was first proposed by Michael Faraday in the early 1800s and became the basis for understanding the relationship between magnetism and the production of electricity.

DMM Terminology . . .

Term	Symbol	Definition
AC		Continually changing current that reverses direction at regular intervals; standard U.S. frequency is 60 Hz
AC COUPLING		Signal that passes an AC signal and blocks a DC signal; used to measure AC signals that are riding on a DC signal
ACCURACY ANALOG METER		Largest allowable error (in percent of full scale) made under normal operating conditions; the reading of a DMM set on the 250 V range with an accuracy rating of ± 2% could vary ± 5 V; analog DMMs have greater accuracy when readings are taken on the upper half of the scale
ACCURACY DIGITAL METER		Largest allowable error (in percent of reading) made under normal operating conditions; a reading of 100.0 V on a DMM with an accuracy of ±2% is between 98.0 V and 102.0 V; accuracy may also include a specified amount of digits (counts) that are added to the basic accuracy rating; for example, an accuracy of ±2% (±2 digits) means that a display reading of 100.0 V on the DMM is between 97.8 V and 102.2 V
AC/DC		Indicates ability to read or operate on alternating and direct current
AC FREQUENCY RESPONSE		Frequency range over which AC voltage measurements are accurate
ALLIGATOR CLIP		Long-jawed, spring-loaded, insulated clamp used to safely make temporary electrical connections
AMBIENT TEMPERATURE		Temperature of air surrounding a DMM or equipment to which the DMM is connected
AMMETER		DMM that measures electric current
AMMETER SHUNT		Low-resistance conductor that is connected in parallel with the terminals of an ammeter to extend the range of current values measured by the ammeter
AMPLITUDE		Measure of AC signal alternation expressed in values such as peak or peak-to-peak
ATTENUATION		Decrease in amplitude of a signal
AUDIBLE	)))))	Sound that can be heard
AUTOHOLD		Function that captures a measurement, beeps, and locks the measurement on the digital display for later viewing and will automatically update with new reading
AUTORANGE MODE		Function that automatically selects a DMM's range based on signals received
AVERAGE VALUE		Value equal to 0.637 times the amplitude of a measured value

. . . DMM Terminology . . .

Term	Symbol	Definition
BACKLIGHT		Light that brightens the DMM display
BANANA JACK	(symbol)	DMM jack that accepts a banana plug
BANANA PLUG	(symbol)	Long, thick terminal connection on one end of a test lead used to make a connection to a DMM
BATTERY SAVE		Feature that enables a DMM to shut down when battery level is too low or no key is pressed within a set time
BNC		Coaxial-type input connector used on some DMMs
CAPTURE		Function that records and displays measured values
CELSIUS	°C	Temperature measured on a scale for which the freezing point of water is 0° and the boiling point is 100°
CLOSED CIRCUIT	(symbol)	Circuit in which two or more points allow a predesigned current to flow
CONTINUITY CAPTURE		Function used to detect intermittent open and short circuits as brief as 250 µs
COUNTS		Unit of measure of DMM resolution; a 1999 count DMM cannot display a measurement of $1/10$ of a volt when measuring 200 V or more; a 3200 count DMM can display a measurement of $1/10$ of a volt up to 320 V
CREST FACTOR		Ratio of peak value to the rms value; the higher the DMM crest factor, the wider the range of waveforms it can measure; in a pure sine wave, the crest factor is 1.41
db READOUT		Decibels (dB) unit of measure used to express the ratio between two quantities such as the gain or loss of amplifiers, filters, or attenuators in telecommunications or audio applications
DC	− − −	Current that constantly flows in one direction
DECIBEL (dB)		Measurement that indicates voltage or power comparison in a logarithmic scale
DIGITS		Indication of the resolution of a DMM; standard DMM with a 3½ digit specification can display three full digits on the right of the display (0 to 9) and ½ digit (1 or left blank) on the left (up to 1999 counts); newer DMMs, such as 6000, 20,000, or 50,000 count DMMs, use counts to more accurately specify resolution
DIODE	(symbol)	Semiconductor that allows current to flow in only one direction
DISCHARGE		Removal of an electric charge
DUAL DISPLAY	(symbol)	Feature that allows two separate measurement parameters to be displayed simultaneously in the meter display
EARTH GROUND		Reference point that is directly connected to ground

. . . DMM Terminology . . .

Term	Symbol	Definition
EFFECTIVE VALUE		Value equal to 0.707 of the peak in a sine wave
FAHRENHEIT	°F	Temperature measured on a scale for which the freezing point of water is 32° and the boiling point is 212°
FREQUENCY		Number of complete cycles occurring per unit of time
FUNCTION SWITCH		Switch that selects the function (AC voltage, DC voltage, etc.) that a DMM is to measure
GLITCH		Momentary spike in a waveform
GLITCH DETECT		Function that increases the DMM sampling rate to maximize the detection of the glitch(es)
GROUND		Common connection to a point in a circuit whose potential is taken as zero
SAVE		Function that allows a measurement to be saved and stored
HARMONICS		Currents generated by electronic devices (nonlinear loads), which draw current in short pulses, not as a smooth sine wave; harmonic currents are whole-number multiples of the fundamental current (typically 60 Hz)
HOLD BUTTON	HOLD H	Button that allows a DMM to capture and hold a stable measurement
INPUT ALERT		Function that provides an audible warning if test leads are in current input jacks but the function switch is not in amps position
LIQUID CRYSTAL DISPLAY (LCD)		Display that uses liquid crystals to display waveforms, measurements, and text on its screen
LoZ		Setting for low-impedance voltages; designed to prevent false ghost voltage readings/displays
MEASURING RANGE		Minimum and maximum quantity that a DMM can safely and accurately measure
MIN MAX		Function that captures and stores the highest and lowest measurements for later viewing; function can be used with any DMM measurement function such as volts, amps, etc.
MIN MAX INSTANTANEOUS PEAK		High-speed response time used to capture MIN MAX readings of a waveform peak value; can be used for crest factor calculations or to capture transient voltage or momentary voltage surge measurements
NOISE		Unwanted extraneous electrical signals
OPEN CIRCUIT		Circuit in which two (or more) points do not provide a path for current flow
OVERLOAD	OL	Condition of a DMM that occurs when a quantity to be measured is greater than the quantity the DMM can safely handle for the DMM range setting or display
PEAK		Maximum value of positive or negative alternation in a sine wave

. . . DMM Terminology

Term	Symbol	Definition
PEAK-TO-PEAK		Value measured from the maximum negative to the maximum positive alternation in a sine wave
POLARITY		Orientation of the positive (+) and negative (–) side of direct current or voltage
PROBE		Pointed metal tip of a test lead used to make contact with the circuit under test
PULSE		Waveform that increases from a constant value, then decreases to its original value
PULSE TRAIN		Repetitive series of pulses
RANGE		Quantities between two points or levels
RECALL		Function that allows stored information (or measurements) to be displayed
RECORD		Allows measurements to be recorded
RESOLUTION		Degree of measurement precision of DMM when taking measurement
RISING SLOPE		Part of a waveform displaying a rise in voltage
ROOT-MEAN-SQUARE		Value equal to 0.707 of the amplitude of a measured value
SAMPLE		Momentary reading taken from an input signal
SAMPLING RATE		Number of readings taken on a signal over time
SHORT CIRCUIT		Two or more points in a circuit that allow an unplanned current flow
TERMINAL		Point to which DMM test leads are connected
TERMINAL VOLTAGE		Voltage level that DMM terminals can safely handle
TREND CAPTURE		Function that allows a measurement to be recorded over time and displayed in a straight line
TRIGGER LEVEL		Fixed level at which DMM counter is triggered
WAVEFORM		Pattern defined by an electrical signal
ZOOM		Function that allows a waveform (or part of waveform) to be magnified

Figure 2-4. DMMs use specific terminology to describe displayed information.

DMM Displays

DIGITAL DISPLAYS

Digital multimeters (DMMs) display readings as exact numerical values. Numerical values displayed digitally on a DMM eliminate errors which can occur when reading an analog meter. Digital display readings show numerical values using a light-emitting diode (LED) or a liquid crystal display (LCD). LED displays are easier to read but use more power than LCDs. Most DMMs have an LCD screen.

Errors can occur when reading a digital display if prefixes, symbols, and decimal points displayed are not properly interpreted. The exact value on a digital display is determined from the number displayed and the position of the decimal point. The selected range determines the placement of the decimal point. For example, voltage ranges on a 50,000 count DMM are 5.0000 V, 50.000 V, and 500.00 V. The DMMs also include very low (50.000 mV and 500.00 mV) and very high (1000 V) ranges. Always check the user's manual for the specific ranges available.

The highest possible reading with the range set on 3 V is 2.999 V. The highest possible reading with the range set on 30 V is 29.99 V. The highest possible reading with the range set on 300 V is 299.9 V. If the range is not high enough, the display reads OL (overload). **See Figure 3-1.** An *autoranging DMM* is a DMM that automatically adjusts to a higher range setting if the range is not high enough to provide the best resolution for the measurement taken.

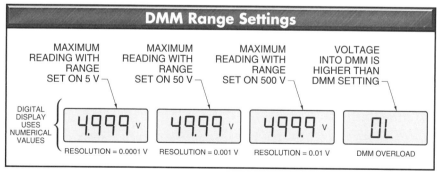

Figure 3-1. The DMM range setting determines the placement of the decimal point.

Bar Graph

Some DMM displays include a bar graph to indicate changes and trends in a circuit. A *bar graph* is a graph composed of individual segments that functions as an analog pointer. **See Figure 3-2.** A negative sign is displayed at the beginning of the bar graph if the polarity of test leads should be reversed. DMMs can have a non-wraparound or wrap-around bar graph.

A bar graph displays the maximum number of segments equal to the DMM range setting. The number of bar graph segments displayed increases as the measured value increases and decreases as the measured value decreases. For example, if the range is set on 30 V, the bar graph displayed has a range of 0 V to 30 V. A display with 23 segments in the 30 V range indicates a 23 V reading.

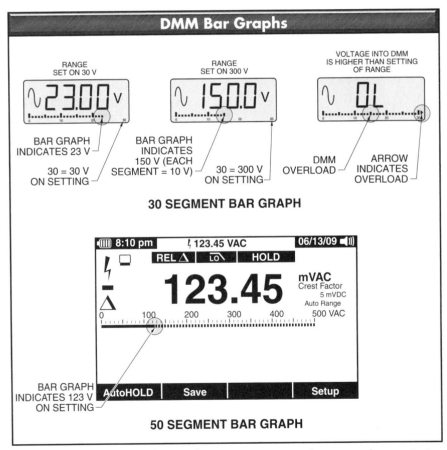

Figure 3-2. DMM bar graphs provide greater sensitivity and accuracy when monitoring measurement changes.

In the 1960s, the integration of transistors on a single crystal of silicon led to the development of entire circuits (integrated circuits) on a tiny semiconductor chip.

A bar graph reading on a display is updated approximately 30 to 40 times per second depending on the DMM type. Generally, 40 times per second is the maximum rate an eye can detect. The numerical reading is updated only about four times per second.

The bar graph is used when quickly changing signals cause the numerical display to flash or when a change in the circuit is too rapid for the numerical display to indicate. For example, mechanical relay contacts may bounce open when exposed to vibration.

Contact bounce causes intermittent problems in electrical equipment. The frequency and severity of contact bounce increases as a relay ages. Resistance changes momentarily from zero to infinity and back when a mechanical relay contact bounces open. A numerical reading on a display cannot indicate contact bounce because most displays require more than 250 ms to update measurements.

A DMM can be used to troubleshoot the 24 VDC electrical system on an airplane.

Contact bounce or loose connections in a circuit are displayed by the movement of one or more segments the moment the contact opens. The quick response of a bar graph more easily enables detection of problems caused by contact bounce or loose connections. In addition, a bar graph indicates measurement changes without the distortion that can occur from reading an analog meter needle at an angle.

Individual segments of the bar graph provide a graphic representation of measurement changes.

GHOST VOLTAGES

DMMs have a high input impedance when set to measure AC voltage. This high input impedance (resistance) ensures that the DMM does not affect the circuit by changing the circuit's resistance (loading the circuit down). Thus, when a DMM is connected to an energized circuit, the circuit is not affected and the measurement is accurate.

However, when a DMM is connected to a de-energized circuit, the high input impedance of the meter can cause voltage to be displayed due to the capacitive coupling between the de-energized parts of the circuit under test and the nearby energized circuits. For example, this may occur when test de-energized conductors are in the same conduit run as energized conductors.

The voltage that appears on the DMM display is not true "power" voltage. It is "static" voltage that does not represent real power (available current). This displayed static voltage is called ghost, or stray, voltage.

Ghost (stray) voltage is a voltage reading on a DMM that is not connected to an energized circuit. For example, if a DMM is connected between a hot wire (energized conductor) and a ground wire that is not connected, ghost voltage can be displayed. **See Figure 3-3.**

Some meters include a low-impedance AC voltage measurement setting. This

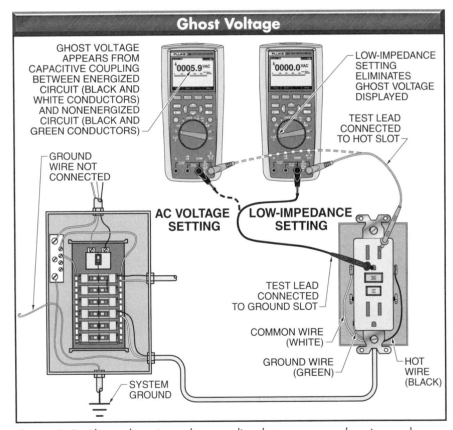

Ghost Voltage

GHOST VOLTAGE APPEARS FROM CAPACITIVE COUPLING BETWEEN ENERGIZED CIRCUIT (BLACK AND WHITE CONDUCTORS) AND NONENERGIZED CIRCUIT (BLACK AND GREEN CONDUCTORS)

GROUND WIRE NOT CONNECTED

LOW-IMPEDANCE SETTING ELIMINATES GHOST VOLTAGE DISPLAYED

TEST LEAD CONNECTED TO HOT SLOT

AC VOLTAGE SETTING

LOW-IMPEDANCE SETTING

TEST LEAD CONNECTED TO GROUND SLOT

COMMON WIRE (WHITE)

GROUND WIRE (GREEN)

HOT WIRE (BLACK)

SYSTEM GROUND

Figure 3-3. Ghost voltage is a voltage reading that appears as changing numbers on the display.

setting is used when testing circuits that may or may not be energized. When taking VAC measurements, always start with the normal VAC meter setting first to ensure the best measurement without placing a load on the circuit. If there appears to be ghost voltage or the measurement does not seem correct, switch the meter to the low-impedance measurement setting.

 For reading accuracy, the next higher range on a DMM display may be selected if the pointer on the bar graph is too sensitive.

The fiber optic source and DMM module accessories are used when troubleshooting fiber optic circuits.

DMM ADVANCED FEATURE APPLICATIONS

A digital multimeter (DMM) is useful when troubleshooting basic voltage, current, and resistance problems. Common troubleshooting problems include testing for power loss from blown fuses, excessive current levels from overloaded circuits, and improper resistance from damaged insulation or equipment. However, when troubleshooting more complex problems, a DMM with advanced features must be used. DMM advanced features are helpful when troubleshooting problems such as improper frequencies, overheated neutrals, and intermittent problems. Advanced features vary with specific DMM models.

Some displays allow different configurations for accurate reading interpretation.

Manual Range Mode and Autorange Mode

Manual Range mode is a DMM mode that allows the selection of a specific measurement range. DMMs have specific ranges for each value measured such as voltage, current, and resistance. For example, a 50,000 count DMM may have six voltage ranges (50.000 mV, 500.00 mV, 5.0000 V, 50.000 V, 500.00 V, and 1000.0 V). A DMM set manually for a specific range can measure values up to the maximum value of the range. For example, a DMM set on the 50.000 V range can display voltage values up to 49.999 V. Values that exceed 50.000 V are overloads and are indicated by OL (overload) on the display. **See Figure 4-1.**

Autorange mode is a DMM mode that automatically selects the range with the best accuracy and resolution for the measurement. The Autorange mode automatically changes the DMM to a higher range if an overload is detected. DMMs usually power up in Autorange mode. The Autorange mode is changed to Manual Range mode with the RANGE button.

A DMM is best left in the Autorange mode when taking most measurements. Using Manual Range mode requires caution as the DMM does not automatically correct for values outside the set range. For example, if the DMM is set to the 4.000 V range, any value exceeding 4.000 V is displayed as OL. The OL reading prevents the display of any measurement over 4.000 V and may lead to confusion regarding the circuit under test.

DMMs are sometimes changed from Autorange mode to Manual Range mode when the value to be measured is known.

For example, a DMM used to measure voltage at a standard 115 VAC residential wall outlet can be set to the 400.0 VAC range to greatly reduce or eliminate ghost voltage. If left in Autorange mode, the DMM defaults during power up to the 400.0 mV range. This lower range is much more sensitive and may result in ghost voltage.

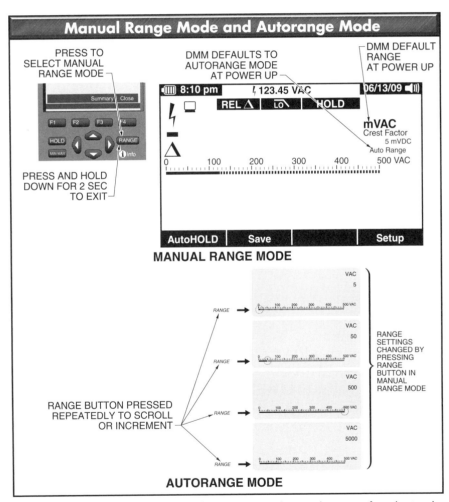

Figure 4-1. Manual Range mode and Autorange mode provide options for selecting the range with the best accuracy and resolution for the measurement.

HOLD Mode

When taking measurements, it is not always convenient to continuously watch the display or see the measurement displayed because of the DMM location. In some cases, the measurement displayed fluctuates rapidly. A DMM with HOLD mode allows a displayed measurement to be saved for reading at a later time or when test leads or attachments are removed from the circuit. **See Figure 4-2.** The two basic types of HOLD modes are the standard HOLD mode and the Auto Hold (Touch Hold®) mode.

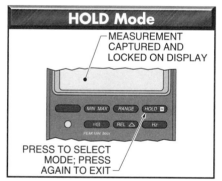

Figure 4-2. HOLD mode captures a measurement for later viewing.

The standard DMM HOLD mode saves ("freezes") the measurement displayed when the HOLD button is pressed. The measurement is retained on the display until the HOLD button is pressed again. The Auto Hold mode retains the measurement displayed when the HOLD button is pressed, but also updates the display when the DMM detects a new stable measurement. The last stable measurement detected is retained on the display until a new stable measurement is detected.

The standard HOLD function is best used when one measurement is to be retained regardless of other measurements detected. The Auto Hold mode is best used for measurements taken in unstable circuits (rapidly changing values) that can cause fluctuating measurements that are difficult to read.

MIN MAX Recording Mode

MIN MAX Recording mode (MIN MAX) is a DMM mode that captures and stores the lowest and highest measurements for later display. MIN MAX Recording mode can be used with DMM values such as voltage, current, and resistance. MIN MAX Recording mode can also be used with accessories to record temperature, pressure, and vacuum measurements. When troubleshooting, MIN MAX Recording mode is especially useful for immediately recording the effect of any adjustments. **See Figure 4-3.**

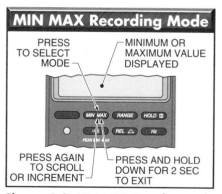

Figure 4-3. MIN MAX Recording mode captures and stores the lowest and highest measurements for later display.

Minimum or maximum values are indicated by pressing the MIN MAX button after the DMM is connected to the circuit. The abbreviation MIN or MAX is displayed with the value to identify the measured value. Most DMMs sound (beep) each time a new minimum or maximum value is recorded. A continuous beep indicates that the DMM is overloaded.

Voltage measurement with MIN MAX Recording mode is useful in identifying low-voltage problems caused by large motors starting up, high-voltage surges, temporary voltage losses, or circuit overloading. For example, most industrial control circuits are powered by a control transformer. The transformer has a fixed power rating (VA or kVA rating) that is listed on the transformer nameplate. If the total load connected to the transformer is within the rating of the transformer, the transformer is delivering the correct secondary voltage. However, if the total load connected to the transformer exceeds the rating of the transformer, the voltage output of the transformer drops. The greater the load, the greater the voltage drop. A voltage drop causes circuit problems such as relays and motor starters dropping out.

Voltage drop is a difficult problem to detect because voltage measurements are often taken when loads in the circuit are OFF, or all loads are not ON. **See Figure 4-4.**

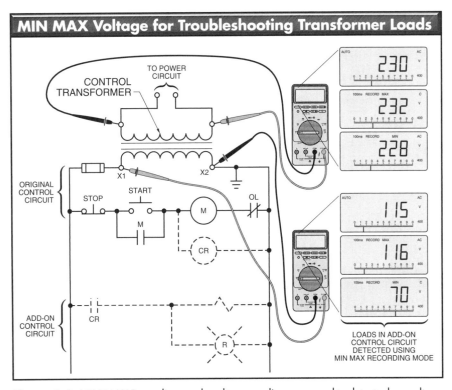

Figure 4-4. MIN MAX Recording mode voltage readings are used to detect a low-voltage condition caused by circuit overloading.

The Institute of Electrical and Electronics Engineers, Inc. (IEEE) is the world's largest professional organization and publishes standards for the electrical/electronics industry.

Because the voltage output of the transformer appears to be within specifications, the true cause of the problem is often overlooked. A DMM set to monitor voltage with MIN MAX Recording mode on all loads during typical system operating conditions can detect a low-voltage condition caused by circuit overloading. The problem can then be corrected by reducing the load on the transformer or increasing the transformer size.

Current measurement with MIN MAX Recording mode is useful in determining if a circuit is overloaded or if there is capacity for additional loads. **See Figure 4-5.** There should be no current drawn with all loads OFF. If there is a current drawn, there may be a leakage current problem. Leakage current problems are caused by a partial short in the system from problems such as insulation breakdown. With all loads ON, the total current drawn should be within the specifications for the system. If the current is less than the maximum circuit capacity, there may be room for additional loads.

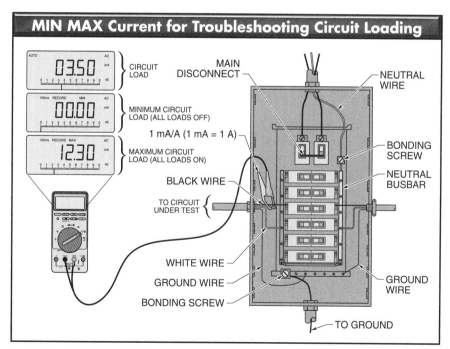

MIN MAX Current for Troubleshooting Circuit Loading

Figure 4-5. MIN MAX Recording mode current readings are used to determine if a circuit is overloaded or if there is capacity for additional loads.

Resistance measurement with MIN MAX Recording mode is useful in isolating problems from loose connections, corrosion, or shorts. **See Figure 4-6.** With all power OFF, the DMM test leads are connected across a connection, splice, load, or circuit. The resistance of the circuit under test is displayed. The DMM is then set to measure resistance using MIN MAX Recording mode, and the wires and/or connections are moved. The measurement should not change.

If a new high-resistance (MAX) measurement is recorded, there is probably a loose connection (open) in the circuit. In a circuit with a fixed-resistance value, such as 200 Ω, a new low-resistance (MIN) measurement indicates there is probably a short or a partial short circuit. A DMM can be left in MIN MAX Recording mode for long-term measurements. The DMM records measurements as long as the battery is good (typically several hundred hours).

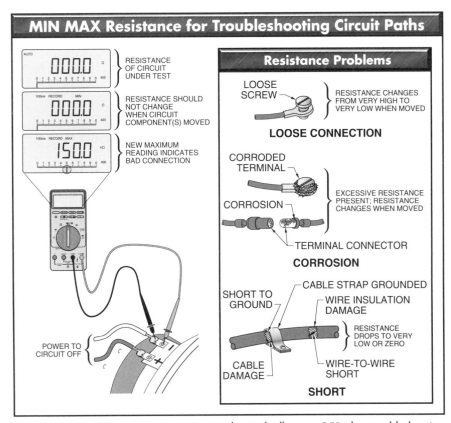

Figure 4-6. Resistance measurements are taken with all power OFF when troubleshooting circuit paths.

Peak MIN MAX Recording Mode

Some DMMs also include a Peak MIN MAX Recording mode. *Peak MIN MAX Recording mode* is a DMM mode that captures maximum readings of AC sine waveform peak values. While in MIN MAX Recording mode, the Peak MIN MAX Recording Mode (or PEAK on some DMMs) button is pressed. A high-speed response time of approximately 250 µsec (microseconds) or less captures the maximum waveform peak value. These readings are useful in crest factor calculations or to capture transient voltages and power surges when using a clamp-on current probe accessory.

Peak MIN MAX Recording mode can also be used to analyze the operation of solid-state motor starters, variable frequency motor drives, computers, and other solid-state electronic devices by the AC sine waveform produced. Normal AC voltages produce a pure AC sine waveform. In a pure AC sine waveform, the peak voltage is equal to 1.41 times the rms voltage.

To obtain these measurements, the DMM is first connected into the circuit for a voltage measurement, and the MIN MAX button is then pressed to record the minimum and maximum rms values. With the DMM in the MIN MAX Recording mode, the Peak MIN MAX Recording Mode button is then pressed. The voltage measurement displayed is the peak voltage. **See Figure 4-7.** For example, with pure AC sine waveform, the peak voltage should be 1.41 times the rms voltage.

A properly operating solid-state motor starter circuit has 232 VAC (rms) and a peak voltage of approximately 327 VAC (232 × 1.41 = 327.12).

If the peak voltage is higher or lower than 1.41 times the rms value, the AC sine waveform is distorted. The greater the difference between the rms and peak values, the greater the AC sine waveform distortion. A meter with a graphic display can be used to further analyze the distorted sine wave.

Average Recording Mode

Some DMMs also include an average value in MIN MAX Recording mode. *Average Recording mode* is a DMM MIN MAX Recording mode that continually calculates the true average of all readings taken over time. This mode is useful for smoothing unstable or changing input measurements and determining average circuit loads.

Minimum and maximum current measurements can be used when troubleshooting the loads on a power distribution system.

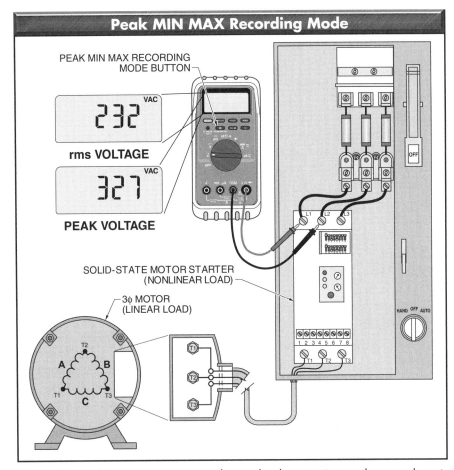

Figure 4-7. A solid-state motor starter with a peak voltage 1.41 times the rms voltage is producing a pure sine waveform.

Relative Mode

Relative mode (REL) is a DMM mode that records a measurement and displays the difference between that reading and subsequent measurements. The Relative mode is typically used to show measurements above or below a specific value, or zero out the baseline reading to eliminate the resistance detected in the test leads. **See Figure 4-8.**

To use Relative mode, the DMM is connected in the circuit. The REL button is pressed and the DMM displays a zero reading. The measured value at the time the REL button is pressed is the stored reference value. In Relative mode, any measurement value displayed on the DMM is the difference between the stored reference value and the measured test value.

Relative Mode

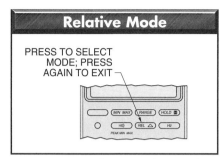

Figure 4-8. Relative mode records a measurement and displays the difference to show measurements above or below a specific value.

Relative mode can be used with measurements such as voltage, current, and resistance. For example, the Relative mode can be used when testing an automotive electrical system for the amount of current drawn by a load or circuit. **See Figure 4-9.** The DMM is set to measure DC current using a clamp-on current probe accessory designed for measuring DC. Current can be measured at the battery for total system current or any individual circuit by measuring at the fuse box.

Relative mode is used to zero out any current that is constantly drawn by the electrical system during normal operation. For example, the memory circuits on computers, digital clocks, and cellular telephones constantly draw a small amount of operating current. Pressing the REL button zeroes out the DMM by subtracting the operating current measurement from the measurement displayed. As an individual load is turned ON, the current measurement displayed is for that load only. This allows greater accuracy in readings taken.

Relative Mode for Measuring Current

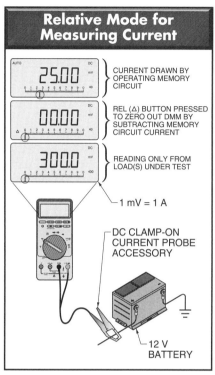

Figure 4-9. Relative mode can be used to zero out the DMM, thereby eliminating the current drawn by memory circuits when measuring individual load current.

The resistance of undamaged test leads should range from 0.2 Ω to 0.5 Ω. The relative mode can be used to zero out test lead resistance.

Input Alert™

Input Alert™ is a constant audible warning emitted by a DMM if test leads are connected in current jacks and a non-current mode is selected. **See Figure 4-10.** Input Alert™ protects against the common mistake of leaving the test leads plugged into the current input jacks and attempting a voltage measurement. This causes a direct short across the source voltage through a low-value resistor, and a high current to flow through the DMM.

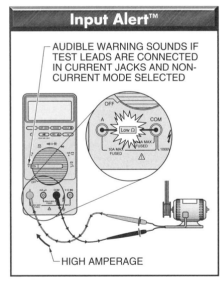

Input Alert™

AUDIBLE WARNING SOUNDS IF TEST LEADS ARE CONNECTED IN CURRENT JACKS AND NON-CURRENT MODE SELECTED

HIGH AMPERAGE

Figure 4-10. Input Alert™ warns the user of incorrect test lead connections.

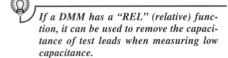 *If a DMM has a "REL" (relative) function, it can be used to remove the capacitance of test leads when measuring low capacitance.*

If the DMM is not adequately protected, the current flow can cause injury to the operator and/or damage to the DMM

and circuit. Input Alert™ helps prevent the nuisance of blowing DMM fuses due to operator error and reduces the risk of damage to the DMM and equipment.

Diode Test Mode

Diode Test mode is a DMM mode used to test diode function. A *diode* is an electronic device that allows current to flow in only one direction. Diodes are commonly used in electrical and electronic circuits to control current flow. In Diode Test mode, the DMM produces a small voltage between the test leads. The voltage is used to test the voltage drop across a diode. Values are then compared with the manufacturer's specifications.

Capacitance Measurement Mode

Capacitance Measurement mode is a DMM mode used to measure capacitance or test a capacitor. A *capacitor* is an electronic device used to store an electrical charge. In Capacitance Measurement mode, the DMM charges the capacitor with a known current, measures the resultant voltage, and calculates the capacitance. Values are then compared with manufacturer's specifications. Capacitance Measurement mode is commonly used to test the condition of capacitors used on capacitor-start motors.

Frequency Counter Mode

Frequency Counter mode is a DMM mode that measures the frequency of AC signals. *Frequency* is the number of cycles per second (cps) in an AC sine wave.

Frequency is measured in hertz (Hz). Values are then compared with system specifications. Frequency measurement is commonly used to test the frequency of AC variable-frequency motor drives.

Duty Cycle Mode

Duty Cycle mode (duty factor) is an alternative Frequency Counter mode that measures the percentage of time a circuit is ON or OFF during a specified period of time. Performance of the circuit and individual loads are then evaluated and compared with operating specifications. Duty Cycle mode is commonly used to test the performance of any pulse-width (on-time) modulated electronic system. For example, on an automotive electronic fuel injection system, pulse width is the length of time in milliseconds that the fuel injector is open to spray fuel.

Event Logging (Recording) Mode

Event Logging (Recording) mode is a DMM mode that records circuit measurement data history over a specific time period. Recording, storing, viewing, and printing out measurements taken with a DMM can be helpful when troubleshooting and/or providing documentation about the operation of a system, circuit, or component. In the Event Logging (Recording) mode, the DMM logs historical data from an input signal (voltage, current, temperature, etc.). The DMM detects each event of stability or instability in the circuit and

then at specified intervals saves information in its memory about that period (start time, stop time, maximum and minimum readings, etc.).

After the data is logged and stored, the data can be viewed. Limited portions of the stored data can be viewed directly on the DMM. In most cases the stored data is downloaded to a personal computer, and software is used to display and analyze the measurement data. The stored data is downloaded to a PC via an infrared adapter. **See Figure 4-11.** Stored data can be displayed in a table and/or in graph form on the PC. The data can then be saved and/or printed as required.

The most common types of diodes include rectifier, zener, tunnel, photoconductive, and light-emitting diodes.

The event logging (recording) mode on a DMM allows measurement data to be downloaded to a PC to be displayed and analyzed.

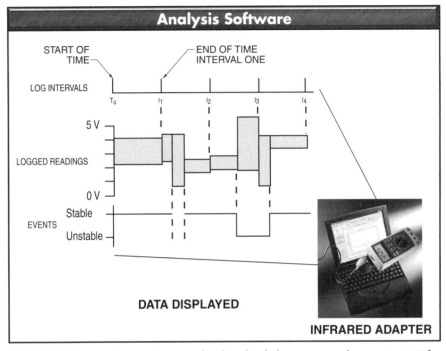

Figure 4-11. Stored event data can be downloaded to a personal computer (PC) for analysis.

AC Low Pass Filter Mode

In some AC circuits there can be more than one frequency. For example, an AC variable-frequency motor drive produces a voltage output to the motor that includes both a carrier frequency and equivalent fundamental frequency produced by the carrier frequency. The carrier frequency is the frequency produced by the drive's internal electronic circuit to produce a simulated power frequency to the motor. This frequency is then varied (changed) to change the speed of the motor. The AC Low Pass Filter mode filters out voltages to accurately measure the resulting composite sine wave. **See Figure 4-12.**

Additionally, the AC Low Pass Filter mode can be used when taking a voltage measurement in any AC circuit that includes (or may include) more than the generated power frequency (60 Hz). This allows a more accurate measurement by filtering out unwanted voltages above a certain frequency.

WARNING: To avoid possible electrical shock, do not use the AC Low Pass Filter mode to verify the presence of hazardous voltages. Voltages greater than indicated may be present. First, take a voltage measurement without using the AC Low Pass Filter mode to detect the possible presence of hazardous voltage. Then, select the AC Low Pass Filter mode as required.

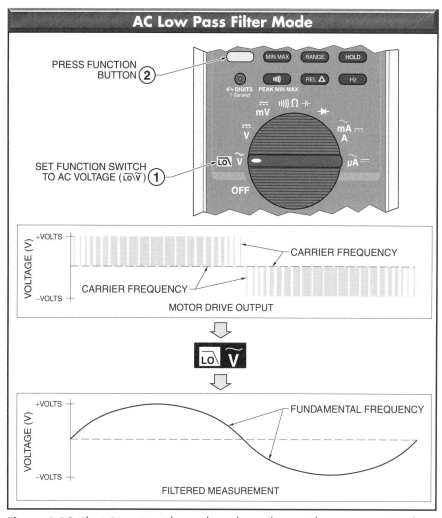

Figure 4-12. The AC Low Pass Filter mode can be used to provide an accurate AC voltage measurement on AC variable frequency motor drives.

Temperature Measurement Mode

Temperature Measurement mode is a DMM mode, with a temperature accessory, that allows direct temperature measurements to be displayed in degrees Fahrenheit or

Celsius. DMM temperature accessories typically output temperature values in millivolts (usually 1 mV per degree) with the meter set to measure millivolts. A DMM with Temperature Measurement mode automatically displays the measurement in degrees to simplify the process.

Measuring AC Voltage

AC VOLTAGE

AC voltage is the most common type of voltage used to produce work. AC voltage is voltage produced by a generator. A *generator* is an electrical device that converts mechanical energy into electrical energy by rotating a wire coil in a magnetic field. An AC generator produces an AC sine wave.

The continuously varying voltage produced by an AC generator is called a sine wave because it follows the sine function. An *AC sine wave* is a symmetrical waveform that contains 360°. An AC sine wave reaches its peak positive value at 90°, returns to zero at 180°, increases to its peak negative value at 270°, and returns to zero at 360°. A *cycle* is one complete wave of alternating voltage or current. A cycle contains 360°. An *alternation* is one half of a cycle. An AC sine wave has one positive alternation and one negative alternation per cycle. **See Figure 5-1.**

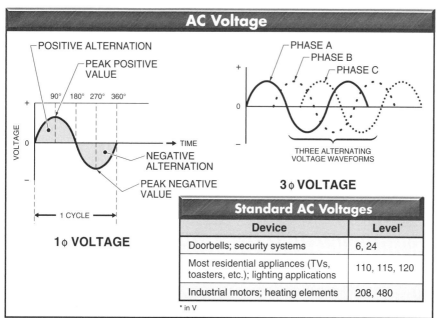

Figure 5-1. A generator converts mechanical energy into electrical energy and produces an AC sine wave.

AC voltage is either single-phase (1φ), or three-phase (3φ). *Single-phase AC voltage* is voltage that contains only one alternating voltage waveform. *Three-phase AC voltage* is voltage that is a combination of three alternating voltage waveforms, each displaced 120° (one-third of a cycle) apart. Three-phase voltage is produced when three coils are simultaneously rotated in a generator.

waveform. Non-sinusoidal waveforms are present in equipment such as variable-speed motor drives, light dimmers, computers, and other solid-state devices. **See Figure 5-2.**

 The international unit of frequency (hertz) is named after Heinrich Hertz who demonstrated the existence of electromagnetic waves (cycles) in 1887.

AC Voltage Measurement

AC voltage can be measured with an average-responding or true-rms DMM. Both DMM types can accurately measure sinusoidal waveforms. A *sinusoidal waveform* is a waveform that is consistent with a pure sine wave. A *non-sinusoidal waveform* is a waveform that has a distorted appearance when compared with a pure sine

AC Voltage Conversions

To Convert	To	Multiply By
rms	Average	.9
rms	Peak	1.414
Average	rms	1.111
Average	Peak	1.567
Peak	rms	.707
Peak	Average	.637
Peak	Peak-to-Peak	2

DMM AC Voltage Measurement

Waveform				
Waveform Type	Sinusoidal		Non-Sinusoidal	
Waveform Shape	Sine Wave	Square Wave	Response to 1φ Distorted Sine Wave	Response to 3φ Distorted Sine Wave
DMM Measurement Accuracy				
Average-Responding DMM	Correct	10% High	40% Low	5% – 30% Low
True-rms DMM	Correct	Correct	Correct	Correct

Figure 5-2. A true-rms DMM is required for measurement accuracy when measuring non-sinusoidal sine wave voltage.

Voltage measurements on sources that operate on non-sinusoidal sine wave voltage require a true-rms DMM for measurement accuracy. AC voltage values are stated and measured as peak, peak-to-peak, average, or rms values. AC voltage values are converted to another value when required.

Peak Voltage Value. The *peak voltage value* (V_{max}) of a sine wave is the maximum value of either the positive or negative alternation. The positive and negative alternation values are equal in a sine wave.

Peak-to-Peak Voltage Value. The *peak-to-peak voltage value* (V_{p-p}) is the value measured from the maximum positive alternation to the maximum negative alternation. **See Figure 5-3.** High-performance DMMs can capture and hold a peak voltage that is present for as little as 1 millisecond (ms). This allows the detection of faults such as power supply surges and voltage spikes.

Average Voltage Value. The *average voltage value* (V_{avg}) of a sine wave is the mathematical mean of all instantaneous voltage values in the sine wave. The average voltage value is equal to .637 of the peak value of a standard sine wave. **See Figure 5-4.** Non-sinusoidal waveforms cause measurement errors in average-responding DMMs. Incorrect measurements are usually lower than actual voltage value.

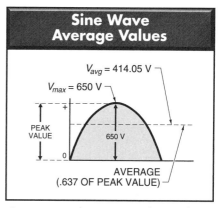

Figure 5-4. The average voltage value is the mathematical mean of all instantaneous voltage values in the sine wave.

AC motors are more widely used than DC motors because most utility companies distribute only AC power.

Root-Mean-Square Voltage Value. The *root-mean-square voltage value* (V_{rms}), or effective value, of a sine wave is the voltage value that produces the same amount of heat in a pure resistive circuit

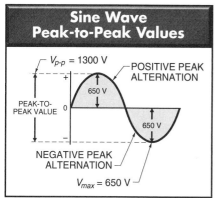

Figure 5-3. High-performance DMMs can quickly capture and hold a peak voltage measurement for detecting power supply surges and voltage spikes.

as DC of the same value. The rms value is equal to .707 of the peak value in a sine wave. **See Figure 5-5.** True-rms DMMs respond accurately to AC voltage values regardless of the waveform and must be used when measuring any non-sinusoidal sine wave voltage.

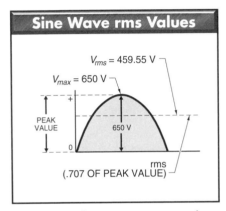

Sine Wave rms Values

V_{rms} = 459.55 V

V_{max} = 650 V

PEAK VALUE

650 V

0

rms
(.707 OF PEAK VALUE)

Figure 5-5. The root-mean-square voltage value of a sine wave produces the same amount of heat in a pure resistive circuit as DC of the same value.

Crest Factor. The *crest factor* is the ratio of the peak voltage value to the rms voltage value. A pure sinusoidal waveform has a 1.41 crest factor. The higher the crest factor in a circuit, the more distorted the waveform. **See Figure 5-6.** In addition, the more distorted the waveform, the harder it is to measure the rms value. Only true rms DMMs have a crest factor rating. The higher the DMM crest factor rating, the greater the range of waveforms that the DMM can accurately measure. A DMM should have a minimum crest factor rating of 3 at full scale and 6 at half scale.

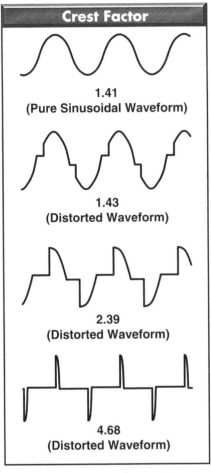

Crest Factor

1.41
(Pure Sinusoidal Waveform)

1.43
(Distorted Waveform)

2.39
(Distorted Waveform)

4.68
(Distorted Waveform)

Figure 5-6. Sine wave distortion is greater as the crest factor increases above 1.41.

AC VOLTAGE MEASUREMENT PROCEDURES

AC voltage measurement procedures may vary slightly with different DMMs. **See Figure 5-7.** Some DMMs have advanced features, which provide convenience and specific measurement capabilities. Always exercise caution with any DMM when measuring AC voltages over 24 V.

WARNING: Ensure that no body part contacts any part of the live circuit, including the metal contact points at the tip of the test leads.

To measure AC voltages with a DMM, apply the procedure:

1. Set the function switch to AC voltage. Set the range to the highest voltage setting if voltage in the circuit is unknown. *Note:* Most DMMs power up in Autorange mode, which automatically selects a measurement range based on voltage present.

2. Plug the black test lead into the common jack.

3. Plug the red test lead into the voltage jack.

4. Connect the black test lead first and red test lead second.

5. Read the voltage measurement displayed. (After measurement is taken, remove the red test lead first and the black test lead second).

Advanced DMM Features

In addition to steps 1–5, the following advanced features included on some DMMs may be used:

6. Press the RANGE button to select a specific fixed measurement range.

7. Press the HOLD button to capture a stable measurement. The measurement recorded can be viewed later.

8. Press the MIN MAX button to capture the lowest or highest measurement. The DMM beeps each time a new reading is recorded.

9. Press the REL button to set the DMM to a specific reference value. Measurements above and below the reference value are displayed.

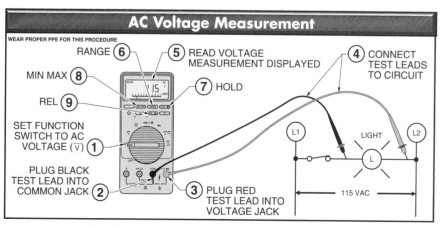

Figure 5-7. Test lead connection to the circuit is arbitrary when measuring AC voltage.

AC Voltage Measurement—Safe Work Practices

A maintenance service call states that there is no power when anything is plugged into a given office receptacle (outlet). As part of the call, voltage measurements are required at the receptacles in the area (Measuring Point 1). Also, both branch circuit voltage and current measurements will be taken in the branch circuit panel (Measuring Point 2) and recorded to determine any future needs.

MINIMUM RECOMMENDATIONS AT MEASURING POINTS:
Point 1 = CAT II Rated DMM and Test Leads
Point 2 = CAT III Rated DMM, Test Leads, and Clamp-on Ammeter
Point 1 = Hazard/Risk Category #1
Point 2 = Hazard/Risk Category #1 ($\leq$ 240 V) and Category #2* (240 V to 600 V)

REQUIRED PPE—POINT 1
* protective long-sleeve shirt rated 4 cal/cm² and pants or coveralls rated 4 cal/cm² (or untreated cotton denim jeans)
* safety glasses
* Class 00 (500 V) rubber insulating gloves with leather protectors
* hard hat

REQUIRED PPE—POINT 2
* protective long-sleeve shirt and pants rated 8 cal/cm²
* safety goggles
* Class 00 (500 V), or Class 0 (1000 V) if voltage is over 500 V, rubber insulating gloves with leather protectors
* wrap-around arc blast face shield (8 cal/cm² minimum rating); if Category #2*, use double-layered switching hood instead of wrap-around arc blast face shield
* protective helmet
* ear protection

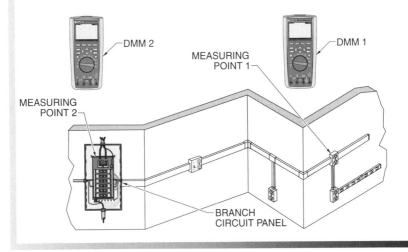

Voltage Measurement Analysis

In general, all AC voltage sources vary from fluctuation in AC voltage over power distribution systems. When different from expected measurement, the voltage is more likely to be lower than normal. In general, voltage measured in AC power systems should be within –10% and +5%. Voltage measurements taken at different points in the system vary. **See Figure 5-8.**

Decibel (dB) Measurements. *A decibel (dB) is the measure of the strength of one electronic signal compared to another.*

The higher the dB number, the greater the signal strength (power gain). A DMM with a dB measurement function can be used to measure the change (gain or loss) in output from an amplifier, signal generator, or other electronic device. **See Figure 5-9.**

A dB measurement requires first taking a measurement at the input and using the Relative (REL) mode to establish a stored reference value of 0 dB. Positive dB values above the 0 dB indicate AC voltages above the reference (gain), and negative dB values indicate AC voltages below the reference (loss).

System Voltage Ranges*

Supply	Service Range		Point of Use Range	
	Satisfactory	Acceptable	Satisfactory	Acceptable
120, 1ϕ	114 – 126	110 – 127	110 – 126	106 – 127
120/240, 1ϕ	114/228 – 126/252	110/220 – 127/254	110/220 – 126/252	106/212 – 127/254
120/208, 3ϕ	114/197 – 126/218	110/191 – 127/220	110/191 – 126/218	106/184 – 127/220
120/240, 3ϕ	114/228 – 126/252	110/220 – 127/254	110/220 – 126/252	106/212 – 127/254
277/480, 3ϕ	263/456 – 291/504	254/440 – 293/508	254/440 – 291/504	254/424 – 293/508

* in volts

Figure 5-8. Voltage measurements vary at different points in the system.

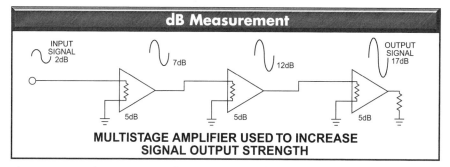

dB Measurement

INPUT SIGNAL 2dB — 7dB — 12dB — OUTPUT SIGNAL 17dB

5dB   5dB   5dB

MULTISTAGE AMPLIFIER USED TO INCREASE SIGNAL OUTPUT STRENGTH

Figure 5-9. A dB measurement function measures the change in signal (gain or loss).

DC VOLTAGE

DC voltage is voltage that flows in one direction only. DC voltage is commonly used in portable equipment such as automobiles, golf carts, flashlights, and cameras. DC voltages power motors, lamps, heaters, and other electrical devices in the same way as AC voltages. The primary difference between AC and DC voltage is the voltage source. AC voltage sources are AC generators. DC voltage sources include voltage directly produced by electromagnetism, chemicals, light, heat, and pressure. **See Figure 6-1.**

Moving molecules have thermal energy. A temperature measurement is a measurement of molecular activity.

In addition to directly produced DC voltage, DC voltage is also obtained from AC voltage passed through a rectifier. A *rectifier* is a device that converts AC voltage to DC voltage by allowing voltage and current to move in one direction only. DC voltage obtained from a rectified AC voltage supply varies from almost pure DC voltage to half-wave DC voltage. **See Figure 6-2.**

DC voltage levels include 1.5 V, 3 V, 6 V, 9 V, 12 V, 24 V, 36 V, 72 V, 90 V, 125 V, 180 V, and 250 V. DC voltage levels from 1.5 V to 72 V are the most common and include automobile and watercraft electrical systems, forklifts, golf carts, and aircraft. DC motors are commonly rated for 6 V, 12 V, 90 V, 120 V, or 180 V. DC voltages exceeding 180 V are used in specialized applications such as small electric railway systems (600 V) and large railway systems (1200 V, 1500 V, and 3000 V).

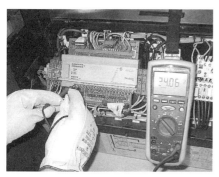
DMMs are commonly used to measure DC voltage when troubleshooting programmable logic controllers (PLCs).

In 1955, the Fluke Model 800A differential DC voltmeter was the first truly portable DC voltmeter with accuracy comparable to laboratory standards.

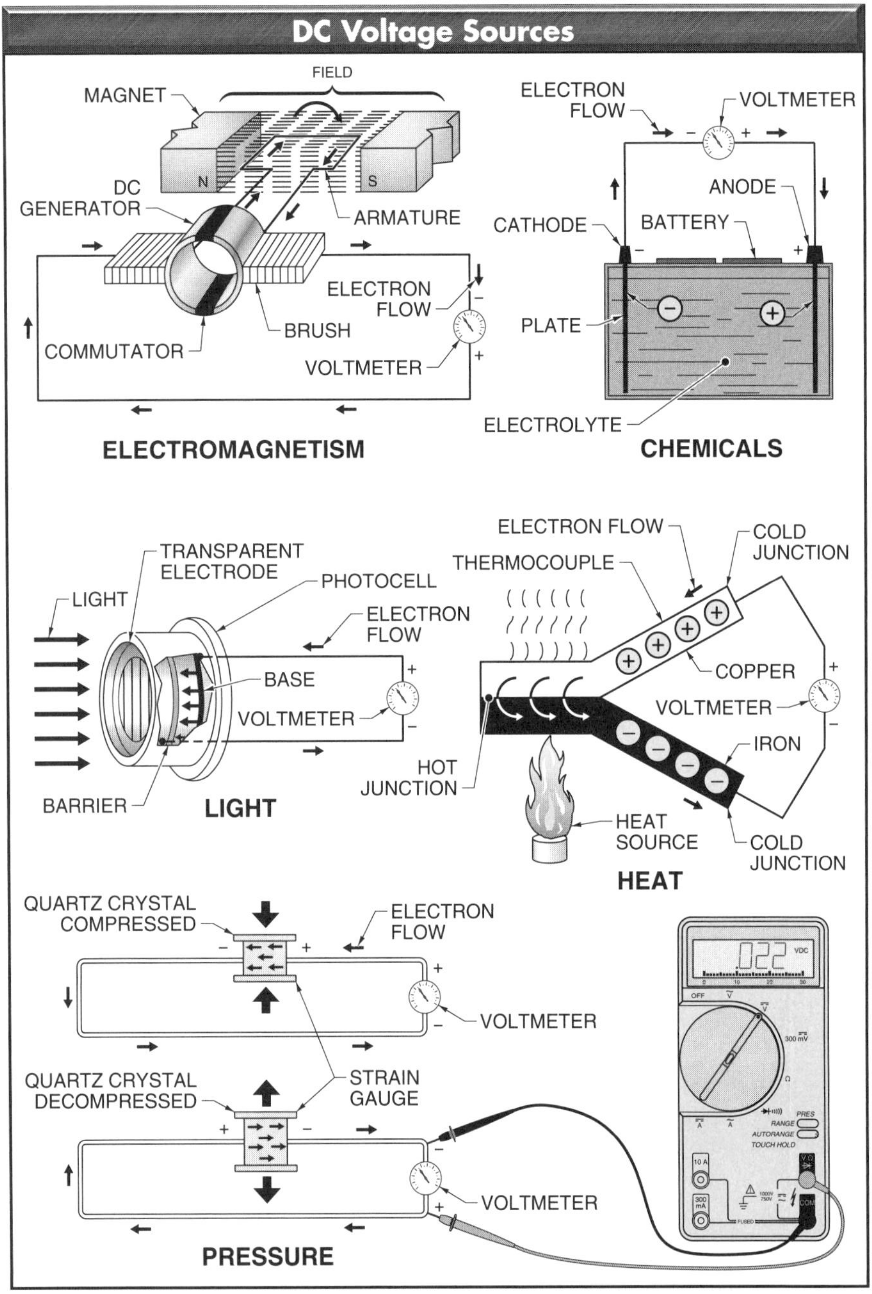

Figure 6-1. DC voltage sources include voltage directly produced by electromagnetism, chemicals, light, heat, and pressure.

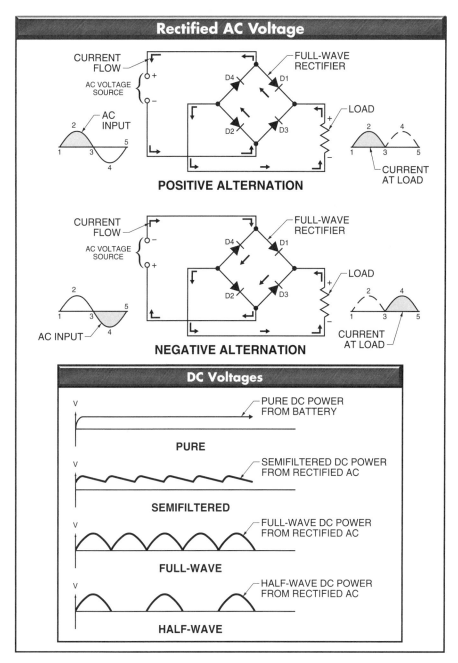

Figure 6-2. DC voltage obtained from a rectified AC voltage supply flows in one direction only.

All DC voltage sources have positive and negative terminals which establish polarity in a circuit. *Polarity* is the positive (+) or negative (−) state of an object. All points in a DC circuit have polarity. If DMM test leads match the polarity of the DC voltage test points (red test lead to positive test point and black test lead to negative test point), a DC voltage measurement is displayed. If DMM test leads do not match the polarity of the DC voltage point being tested, (red test lead to negative test point and black test lead to positive test point), a negative sign appears to the left of the DC voltage measurement displayed. **See Figure 6-3.**

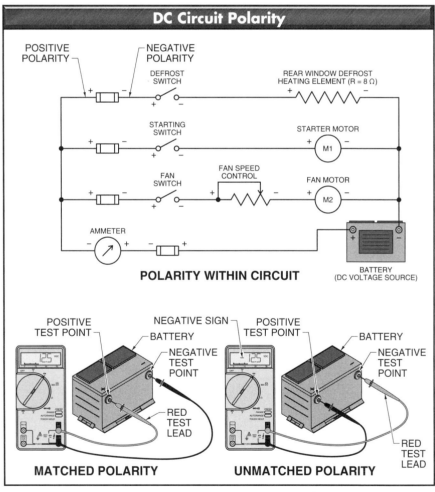

Figure 6-3. All DC circuits have polarity throughout the circuit. If the polarity is unmatched with the DMM test leads, a negative sign appears on the display.

DC VOLTAGE MEASUREMENT PROCEDURES

Exercise caution when taking any circuit measurement. Measurements of DC voltages exceeding 60 V and DC measurements near any battery require extra caution.

WARNING: When testing larger batteries, such as an automobile battery, make sure the test area is clear of any objects that could cause a short. A battery can be shorted by contact with a metal object between the positive and negative battery terminals. Automobile batteries deliver a large amount of power and, when shorted, can explode.

DC voltage is measured with a DMM using standard procedures. **See Figure 6-4.**

1. Set the function switch to DC voltage. If the DMM includes more than one DC setting, select the highest setting. For example, some DMMs include a VDC setting and an mVDC setting. The VDC setting should be used.

2. Plug the black test lead into the common jack.

3. Plug the red test lead into the voltage jack.

 With two batteries, voltage increases when connected in series and current increases when connected in parallel.

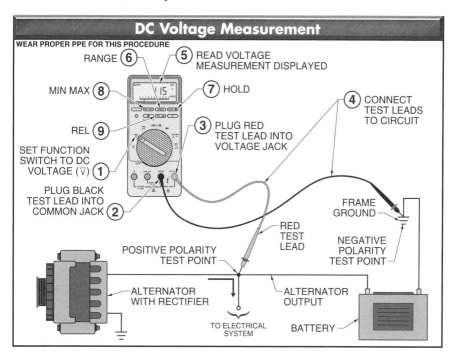

Figure 6-4. DC voltage measurements using a DMM are taken by connecting the black test lead to the negative polarity test point and connecting the red test lead to the positive polarity test point.

4. Connect the test leads to the circuit. The black test lead is connected to the negative polarity test point (circuit ground) and the red test lead to the positive polarity test point. Reverse the test leads if a negative sign (–) appears to the left of the measurement displayed.

5. Read voltage measurement displayed.

Advanced DMM Features

In addition to steps 1 – 5, the following advanced features included on some DMMs may be used:

6. Press the RANGE button to select a specific fixed measurement range. If the voltage measurement is within the range of a lower VDC setting, such as the mVDC, a more accurate measurement can be obtained by changing the setting. Disconnect the

positive (red) test lead before disconnecting the negative (black) test lead in the circuit. Change the DMM setting, reconnect the DMM back into the circuit at the same test points, and read the measurement displayed.

7. Press the HOLD button to capture a stable measurement. The measurement recorded may be viewed later.

8. Press the MIN MAX button to capture the lowest and highest measurement. The DMM beeps each time a new reading is recorded.

9. Press the REL (relative) button to set the DMM to a specific reference value. Measurements above and below the reference value are displayed.

The voltage range for electrical and electronic equipment is +5% to –10% because overvoltage is generally more damaging than undervoltage.

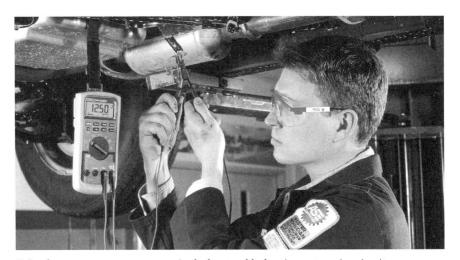

DC voltage measurements are required when troubleshooting automotive circuits.

DC Voltage Measurement—Safe Work Practices

A maintenance service call states that a hoist motor seems to have lost some power. Since the 180 VDC hoist motor is not easily accessible, a check at the hoist control module in the motor control center (Measuring Point 1) will be made first. A 180 VDC measurement is taken at the connection points out of the DC drive/starter to verify the motor is receiving full voltage since a lower DC voltage will reduce the motor torque and speed. If all measurements at the motor control center are good, a measurement is taken directly at the motor (Measuring Point 2).

MINIMUM RECOMMENDATIONS AT MEASURING POINTS:

Point 1 = CAT III Rated DMM and Test Leads

Point 2 = CAT II Rated DMM and Test Leads

Point 1 = Hazard/Risk Category #2*

Point 2 = Hazard/Risk Category #1

REQUIRED PPE—POINT 1

- protective long-sleeve shirt and pants rated 8 cal/cm²
- safety goggles
- Class 00 (500 V) rubber insulating gloves with leather protectors
- double-layered switching hood (8 cal/cm² minimum rating)
- protective helmet
- ear protection

REQUIRED PPE—POINT 2

- protective long-sleeve shirt rated 4 cal/cm² and pants or coveralls rated 4 cal/cm² (or untreated cotton denim jeans)
- safety glasses
- Class 00 (500 V) rubber insulating gloves with leather protectors
- hard hat

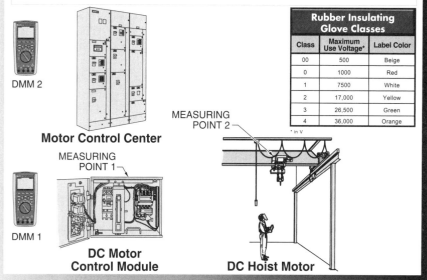

Rubber Insulating Glove Classes		
Class	Maximum Use Voltage*	Label Color
00	500	Beige
0	1000	Red
1	7500	White
2	17,000	Yellow
3	26,500	Green
4	36,000	Orange

* in V

DMM 2

Motor Control Center

MEASURING POINT 1

DMM 1

DC Motor Control Module

MEASURING POINT 2

DC Hoist Motor

Voltage Measurement Analysis

Proper DMM settings and connections are required for obtaining accurate voltage measurements. However, voltage measurements are only the first step in testing or troubleshooting a circuit. Analyzing the voltage measurements is the second step. Unlike AC voltages that can generally vary between −10% and +5% of the power source rating without causing any problems, even small DC voltage variations may indicate a problem.

The exact amount of acceptable DC voltage variation depends upon the application. **See Figure 6-5.** For example, a fully charged, 12 V rated automobile battery may have an open-circuit voltage ranging from 11.9 V to 12.6 V. An 11.9 V measurement indicates a dead battery, a 12.0 V measurement indicates approximately a 25% charge, a 12.2 V measurement indicates approximately a 50% charge, a 12.4 V measurement indicates approximately a 75% charge, and a 12.6 V measurement indicates approximately a fully charged battery. An automobile battery rated at 12 V with a slightly higher voltage measurement (approximately 3% to 5%) is much better than a slightly lower voltage measurement. A DC voltage variation below the normal rated voltage in an automobile battery indicates a problem.

There are some DC applications in which large DC voltage variations are not only acceptable, but intentional. For example, the speed of DC motors can be adjusted by varying the amount of DC voltage supplied. In these applications, the DC motor voltage measured depends on the setting of the voltage regulator. When taking and comparing DC voltage measurements, manufacturer's specifications should be referenced for specific values in the circuit.

Electronic circuits in computers, audio equipment, home appliances, and security systems are powered by DC voltage that has been rectified from AC voltage from a wall outlet.

Battery Charge		
Voltage*	Charge†	State
11.9	0	Dead
12.0–12.1	25	Minimal Charge
12.2–12.3	50	Half Charge
12.4–12.5	75	Slight Undercharge
12.6	100	Full Charge

* in V
† in %

Figure 6-5. Twelve volt batteries range from 11.9 V (dead) to 12.6 V (fully charged).

In some applications, DC voltage measurements may be taken in circuits that include AC voltage. For maximum accuracy of the DC voltage measurement, the AC voltage is measured first and recorded. The DC voltage is then measured by selecting a DC voltage range with the RANGE button that is the same or higher than the AC voltage range.

Both AC and DC Voltage Measurements

In some cases, a signal may contain both an AC voltage and a DC voltage. Some DMMs can simultaneously measure and display both the AC and DC components of a signal separately or as a combined AC + DC value (rms value). **See Figure 6-6.**

The AC part of the signal can be shown on the primary display and the DC part of the signal is shown on the secondary display. The display settings can usually be changed so that the DC part of the signal is shown on the primary display and the AC part of the signal on the secondary display, or the displayed measurement can be the equivalent rms signal value (AC + DC value).

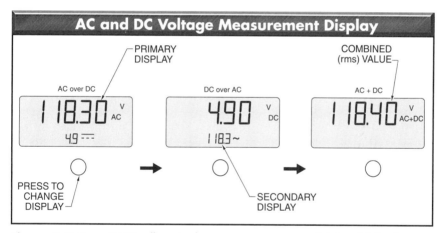

Figure 6-6. Some DMMs allow simultaneous measurements of AC and DC voltages.

Measuring Resistance and Continuity Testing

DIGITAL MULTIMETER PRINCIPLES

RESISTANCE

Resistance (R) is the opposition to the flow of electrons in a circuit. Resistance is measured in ohms (Ω). Higher resistance measurements are expressed using prefixes, as in kilohms (kΩ) and megohms (MΩ). Resistance measurements on a DMM are displayed as Ω, kΩ, and MΩ. **See Figure 7-1.** Prefixes used simplify the measurement displayed. **See Appendix.**

to supply voltage to the test leads and component under test. All resistance measurements must be taken with the circuit de-energized. If a circuit includes a capacitor, the capacitor must be discharged before taking any resistance reading.

A DMM is used to take measurments when troubleshooting telephone wiring systems.

Figure 7-1. DMMs use electrical prefixes with higher resistance measurements to simplify the measurement displayed.

DMMs measure the amount of resistance in a component or circuit that is not energized. Power is not required because the DMM has an internal battery used

Resistance measurements are normally taken to indicate the condition of a component or circuit. The higher the resistance, the lower the current flow. Likewise, the lower the resistance, the higher the current flow. Components designed to insulate, such as rubber or plastic, should have a

Resistance Readings

AUTO
10.00 kΩ
0 1 2 3 4 5 6 7 8 9 0   40
+

= 10,000 Ω

AUTO
0.001 MΩ
0 1 2 3 4 5 6 7 8 9 0   4
+

DMM USES PREFIX ON HIGHER RESISTANCE READINGS

very high resistance. Components designed to conduct, such as conductors or switch contacts, should have a very low resistance. When insulators are damaged by moisture and/or overheating, resistance decreases. When conductors are damaged by burning and/or corrosion, resistance increases. Other components such as heating elements and resistors should have a fixed-resistance value. Any significant change in the fixed-resistance value usually indicates a problem. Some components, such as resistors, include a resistance value on the component. When a tolerance is indicated, the measured resistance value should be within the specified resistance range. **See Figure 7-2.**

 Copper has a lower resistance than aluminum and is the most common material used for conductors.

DMM Resistance Settings

On most DMMs, the Resistance mode shares the function switch position with another DMM mode. Depending on the DMM used, the Resistance mode may also include the Continuity Test mode, the Capacitance Measurement mode, and/or the Diode Test mode. **See Figure 7-3.** Since resistance measurements are the most common of these modes, DMMs commonly power up in the Resistance mode.

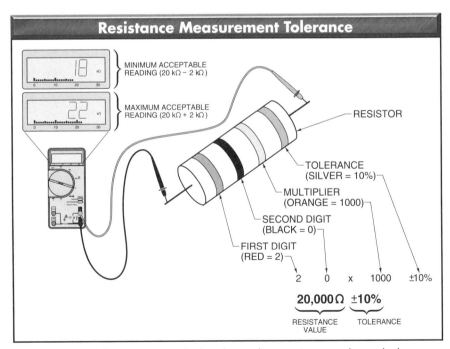

Figure 7-2. Small resistors use color bands to indicate resistance value and tolerance.

Resistance Mode Function Switch Positions

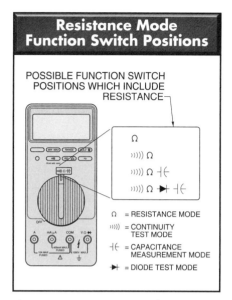

Figure 7-3. On most DMMs, the Resistance mode function switch position is shared with other DMM modes.

A DMM in Resistance mode automatically begins to take a resistance measurement, even before the test leads are connected to the component tested. This results in the DMM display indicating OL. The OL reading also includes the MΩ symbol because resistance of the open test leads (not connected together or to the test component) is extremely high.

When the test leads are connected to the test component, the DMM automatically adjusts by Autorange mode to the best range. The range can be manually set to any of the DMM resistance ranges. When the RANGE button is pushed and the test leads are open, the decimal point on the OL displayed changes (.OL, O.L, OL.), and the displayed prefix changes

(MΩ, kΩ). On most DMMs, the range value (4, 40, 400) on the bar graph changes. **See Figure 7-4.**

Resistance Range Values

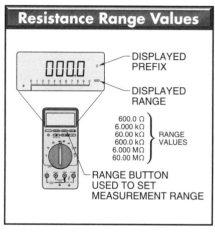

Figure 7-4. The RANGE button is used to manually set the DMM to a specific resistance range.

RESISTANCE MEASUREMENT PROCEDURES

A DMM measures resistance with the circuit or component de-energized. Low voltage applied to a DMM set to measure resistance causes inaccurate readings. High voltage applied to a DMM set to measure resistance causes DMM damage, even if the DMM has internal protection. Always check for voltage before taking any resistance measurements in a circuit.

Resistance is measured with a DMM using standard procedures. **See Figure 7-5.**

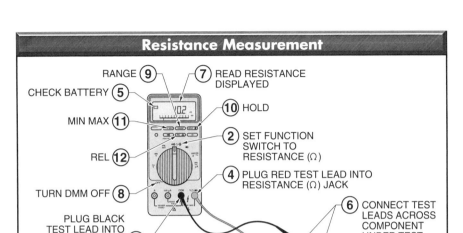

Figure 7-5. Resistance measurements are taken with the circuit de-energized.

1. Check that all power is OFF to the circuit and/or remove component under test.

2. Set the function switch to Resistance mode as required on the DMM. The DMM should display OL and the Ω symbol when the DMM is in the Resistance mode.

3. Plug the black test lead into the common jack.

4. Plug the red test lead into the resistance jack.

5. Ensure that the DMM batteries are in good condition. The battery symbol is displayed when the batteries are low.

6. Connect the test leads across the component under test. Ensure that contact between the test leads and the circuit is good. If the test leads are shorted together, the DMM reading should be 0 Ω. However, the resistance of the test leads (typically 0.2 Ω to 0.5 Ω) may be displayed and can affect a very low-resistance measurement. In very low-resistance measurements, the Relative mode should be used to automatically subtract test lead resistance from any resistance measurement taken. Dirt, solder flux, oil, and other foreign substances can greatly affect resistance readings.

Likewise, body contact with the metal ends of the test leads also affects the resistance measurement. This occurs because resistance through the body becomes a parallel resistance path lowering the total circuit resistance.

7. Read the resistance displayed on the DMM.

8. After completing all resistance measurements, turn the DMM OFF to prevent possible battery drain.

Advanced DMM Features

In addition to steps 1 – 8, the following advanced features included on some DMMs may be used:

9. Press the RANGE button to select a specific fixed measurement range.

10. Press the HOLD button to capture a stable measurement. The measurement recorded can be viewed later.

11. Press the MIN MAX button to capture the lowest and highest measurement. The DMM beeps each time a new reading is recorded.

12. Press the REL button to set the DMM to a specific value. Measurements above and below the reference value are displayed.

Resistance Measurement Analysis

The significance of a resistance reading depends on the component under test. In general, resistance of any one component

varies over time, and from component to component. Slight resistance changes are usually not critical but may indicate a pattern that should be noted.

For example, as the resistance of a heating element increases, the current passing through the heating element decreases, and the power produced by the heating element also decreases. Likewise, as the resistance of a heating element decreases, current passing through the heating element increases, and the power produced by the heating element also increases. In general, the resistance of components used to control circuits (switches, relay contacts, etc.) starts out very low and increases over time from causes such as wear and dirt. Loads such as motors and solenoids decrease in resistance over time from insulation breakdown and moisture.

The resistance measurement displayed by the DMM is the total resistance through all possible paths between the test leads. Caution is required when taking resistance measurements across a component that is part of a circuit. **See Figure 7-6.** The resistance of all components connected in parallel with the component tested affects (usually lowers) the resistance reading. Always check the circuit schematic for parallel paths. In addition, touching the exposed metal parts of the test leads during the test can also cause reading errors.

In 1949, the Fluke Model 301A was introduced as the industry's first super-regulated, high-power, DC current source. The 301 Series power supplies were widely used in the early phases of nuclear weapons testing.

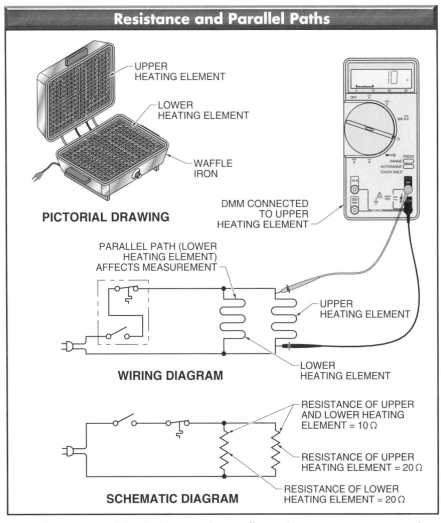

Figure 7-6. Any parallel path in the circuit has an effect on the resistance measurement taken.

CONTINUITY

Continuity is the presence of a complete path for current flow. For example, a closed switch that is good has continuity. An open switch does not have continuity. The Continuity Test mode on a DMM

can be used to test components such as switches, fuses, electrical connections, and individual conductors. The DMM emits an audible response (beeps) when there is a complete path. Indication of a complete path can be used to determine the condition of a component as open or closed. For example, a good fuse should have continuity, whereas a bad fuse does not have continuity.

The main advantage of using the Continuity Test mode is that an audible response is sometimes more desirable than reading a resistance measurement. An audible response allows the technician to concentrate on the testing procedures without looking at the DMM display. When testing for continuity, the DMM beeps based on the resistance of the component under test.

The resistance is determined by the range setting of the DMM. **See Figure 7-7.** For example, if the range is set to 400.0 Ω, the DMM typically beeps if the component under test has a resistance of 40 Ω or less.

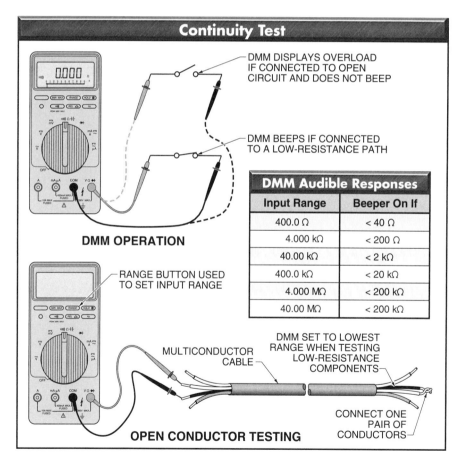

Continuity Test

DMM DISPLAYS OVERLOAD IF CONNECTED TO OPEN CIRCUIT AND DOES NOT BEEP

DMM BEEPS IF CONNECTED TO A LOW-RESISTANCE PATH

DMM OPERATION

RANGE BUTTON USED TO SET INPUT RANGE

DMM Audible Responses

Input Range	Beeper On If
400.0 Ω	< 40 Ω
4.000 kΩ	< 200 Ω
40.00 kΩ	< 2 kΩ
400.0 kΩ	< 20 kΩ
4.000 MΩ	< 200 kΩ
40.00 MΩ	< 200 kΩ

MULTICONDUCTOR CABLE

DMM SET TO LOWEST RANGE WHEN TESTING LOW-RESISTANCE COMPONENTS

CONNECT ONE PAIR OF CONDUCTORS

OPEN CONDUCTOR TESTING

Figure 7-7. Audible responses allow testing for continuity without viewing the DMM display. The range setting determines the resistance at which the DMM indicates continuity.

If the range is set to 4.000 MΩ, the DMM typically beeps if the component under test has a resistance of 200 kΩ or less. The lowest range setting should be used when testing circuit components that should have a low-resistance value, such as electrical connections or switch contacts.

CONTINUITY MEASUREMENT PROCEDURES

Continuity is tested with a DMM using standard procedures. **See Figure 7-8.**

1. Set the function switch to Continuity Test mode as required on the DMM. On most DMMs, the Continuity Test mode and Resistance mode share the same function switch position. The sign should appear in the DMM

display. The DMM may still display OL and Ω.

2. Press the continuity button if required.

3. Plug the black test lead into the common jack.

4. Plug the red test lead into the resistance jack.

5. With the circuit de-energized, connect the test leads across the component under test. The position of the test leads is arbitrary.

6. If there is a complete path (continuity), the DMM beeps. If there is no continuity (open circuit), the DMM does not beep.

7. After completing all continuity tests, turn the DMM OFF to prevent possible battery drain.

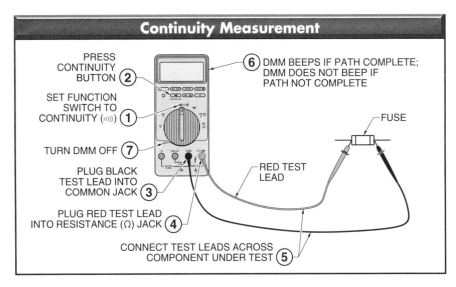

Figure 7-8. The DMM beeps if there is continuity in the component or circuit.

CURRENT

Current is the amount of electrons flowing through an electrical circuit. Current is measured in amperes. An *ampere* is the number of electrons passing a given point in one second. Large amounts of current are measured in amperes (A). Small amounts of current are measured in milliamperes (mA) or microamperes (µA). Current flows through a circuit when a power source is connected to a device that uses electricity. **See Figure 8-1.**

Amperage measurements are taken to determine the amount of circuit loading or the condition of an electrical component (load). Every load (lamp, motor, heating element, speaker, etc.) that converts electrical energy into some other form of energy (light, rotating motion, heat, sound, etc.) uses current. The more electrical energy required, the higher the current usage. Every time a new load is added to a circuit (or switched ON), the circuit must provide more current.

Currents and Current Levels

Like voltage, current may be direct or alternating. *Direct current (DC)* is current that flows in one direction only. Direct current flows in any circuit connected to a power source producing DC voltage, such as a battery or DC generator. *Alternating current (AC)* is current that reverses its direction of flow at regular intervals. Alternating current flows in any circuit connected to a power source producing AC voltage. AC generators produce an AC voltage, and transformers are used to increase (step up) or decrease (step down) AC voltages.

Amperage measurements can be used to assess the condition of the load(s) in a circuit.

Different DC power sources produce different amounts of current. For example, standard size AAA, AA, C, and D batteries all produce 1.5 V, but each size is capable of delivering different amounts of current.

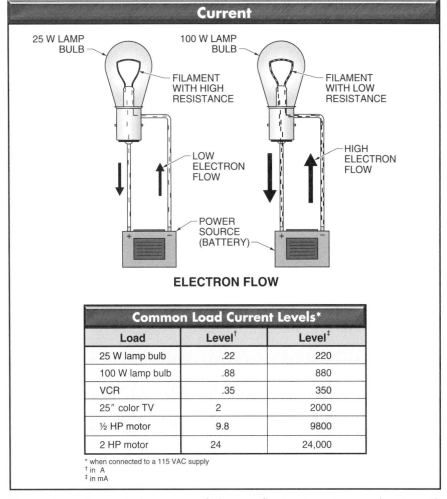

Common Load Current Levels*		
Load	Level[†]	Level[‡]
25 W lamp bulb	.22	220
100 W lamp bulb	.88	880
VCR	.35	350
25" color TV	2	2000
½ HP motor	9.8	9800
2 HP motor	24	24,000

* when connected to a 115 VAC supply
[†] in A
[‡] in mA

Figure 8-1. Current is the amount of electrons flowing in a circuit and varies with the application.

AAA batteries are capable of delivering the least amount of current, and D batteries are capable of delivering the most amount of current. A load connected to a D battery is energized longer than the same load connected to a AAA battery. To increase the available amount of current in a circuit, DC power sources are connected in parallel. To increase the available amount of voltage in a circuit, power sources are connected in series. **See Figure 8-2.**

Current limits are based on the rated load current limit of electrical components such as conductors, switches, motors, and transformers in the circuit. *Overcurrent* is a condition that exists in an electrical circuit when the normal load current is exceeded. When the normal load current is exceeded, excessive heat is produced. The longer and higher the overcurrent, the greater the amount of heat produced. Overcurrent heat can cause insulation to break down, components to fail, and a possible fire hazard. To prevent overcurrents, current limits are set by the proper sizing of fuses, circuit breakers, overloads, and other current-monitoring devices.

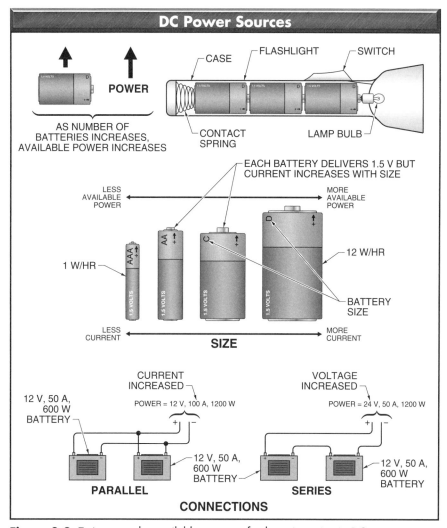

Figure 8-2. To increase the available amount of voltage in a circuit, DC power sources are connected in series.

DC Flow

Early scientists believed that electrons flowed from positive (+) to negative (–). Later, when atomic structure was studied, electron flow from negative to positive was introduced. *Conventional current flow* is current flow from positive to negative. *Electron current flow* is current flow from negative to positive.

Both conventional current flow and electron current flow theories are still used to describe how current flows in a circuit. Conventional current flow is commonly used in automotive electrical circuits. Electron current flow is commonly used in solid-state electronic circuits. **See Figure 8-3.**

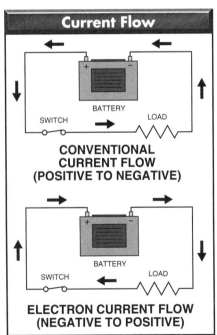

Current Flow

CONVENTIONAL CURRENT FLOW (POSITIVE TO NEGATIVE)

ELECTRON CURRENT FLOW (NEGATIVE TO POSITIVE)

Figure 8-3. Electron flow direction in a circuit is described by the conventional current flow and electron current flow theories.

CURRENT MEASUREMENT PROCEDURES

Current is measured using a clamp-on ammeter, a DMM with a clamp-on current probe accessory, or a DMM connected as an in-line ammeter. **See Figure 8-4.** A *clamp-on ammeter* is a meter that measures current in a circuit by measuring the strength of the magnetic field around a single conductor. The clamp-on ammeter's jaws are placed around the conductor to measure current. Clamp-on ammeters are available for measuring AC only, or for measuring both AC and DC. Clamp-on ammeters commonly include features that also allow for the measurement of AC voltage, DC voltage, and resistance.

When a clamp-on current probe accessory is used with a DMM, the DMM functions as a clamp-on ammeter. Clamp-on current probe accessories are available for DMMs that allow for the measurement of AC only or for the measurement of AC and DC. A DMM connected as an in-line ammeter measures current in a circuit by inserting the DMM in series with the component(s) under test. In-line ammeter readings require the circuit to be opened so the ammeter can be inserted in the circuit. In-line ammeters are available for measuring DC only, or AC and DC.

If voltage in a circuit remains constant, current decreases with an increase in resistance, and current increases with a decrease in resistance.

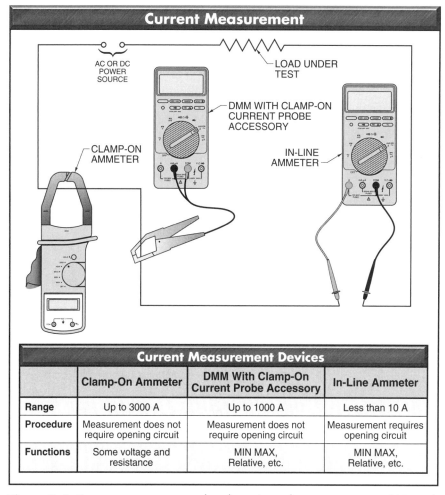

Figure 8-4. Current measurements can be taken using a clamp-on ammeter, a DMM with a clamp-on current probe accessory, or an in-line ammeter.

The advantage of a clamp-on ammeter or a DMM with a clamp-on current probe accessory is that readings can be taken more safely without opening or interrupting the circuit. Both are commonly used to measure currents from 1 A (or less) to 3000 A. The main advantage of using a DMM with a clamp-on current probe accessory instead of a clamp-on ammeter is the features available on the DMM such as MIN MAX Recording mode, Relative mode, and bar graph display. These additional features can be very helpful when troubleshooting. Some DMMs also measure rms currents, as in rms voltage measurements.

In-line ammeters are used to measure very small amounts of current normally less than 10 A. The maximum current that can be measured is determined by the function switch position (A, mA, and μA) and the current rating listed on the current jacks (10 A, 400 mA, etc.). Exceeding the current rating listed on the current jacks could cause injury, damage the meter (unfused current jacks), or blow the fuse (fused current jacks). A DMM typically has two current jacks. One jack is used for measuring lower currents (typically up to 400 mA). The other jack is used for measuring higher currents (typically up to 10 A). In-line current measurements should be limited to circuits that can be easily opened and circuits known to have currents less than 10 A.

WARNING: In-line current measurements are taken with the circuit energized. Proper safety precautions and PPE must ensure that no part of the body comes in contact with a live part of the circuit.

Generally, the first choice for measuring current in a circuit should be a clamp-on ammeter or a DMM with a clamp-on current probe accessory. If a clamp-on ammeter cannot be used, or an in-line measurement is required, steps must be taken to prevent possible injury and/or damage to equipment.

Fuses and circuit breakers are connected in series with circuit conductors. No current flows in any part of a series circuit when a fuse blows or a breaker opens. Check the fuses or breakers first if a circuit has no power.

Testing DMM Fuses

Most DMMs have fuses that become part of the current-measuring circuit. The fuses help to prevent possible injury and/or equipment damage. The DMM fuses must be tested before taking any current measurements. A blown (open) DMM fuse prevents the measurement of any circuit current, even if there is current in the circuit.

Most DMM manufacturers specify a fuse-testing procedure. If a fuse-testing procedure is not provided, DMM fuses must be tested by using the DMM on a low-current test circuit known to be good. A typical fuse-testing procedure provided by the manufacturer lists basic steps. **See Figure 8-5.**

1. Set the function switch to resistance.

2. Plug the red test lead into the voltage jack.

3. Touch the red test lead tip in the DMM high-current terminal (A). Move the test lead tip of the test lead probe around until a reading is obtained.

4. A resistance reading (typically 0.0 Ω to 0.5 Ω) should be displayed. The exact reading should be verified with the user's manual. If the DMM reads OL, replace the fuse and test again. If the DMM reads any other resistance value, refer to the user's manual or have the DMM serviced.

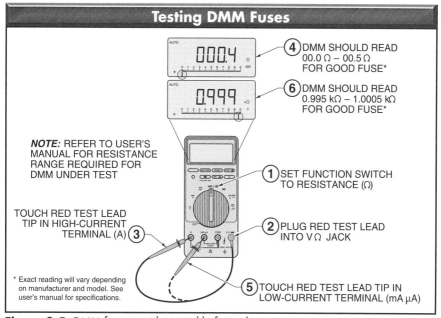

Testing DMM Fuses

④ DMM SHOULD READ
00.0 Ω – 00.5 Ω
FOR GOOD FUSE*

⑥ DMM SHOULD READ
0.995 kΩ – 1.0005 kΩ
FOR GOOD FUSE*

NOTE: REFER TO USER'S
MANUAL FOR RESISTANCE
RANGE REQUIRED FOR
DMM UNDER TEST

① SET FUNCTION SWITCH
TO RESISTANCE (Ω)

TOUCH RED TEST LEAD
TIP IN HIGH-CURRENT
TERMINAL (A) ③

② PLUG RED TEST LEAD
INTO V Ω JACK

* Exact reading will vary depending
on manufacturer and model. See
user's manual for specifications.

⑤ TOUCH RED TEST LEAD TIP IN
LOW-CURRENT TERMINAL (mA µA)

Figure 8-5. DMM fuses must be tested before taking current measurements.

5. Touch the red test lead tip in the DMM low-current terminal (mA µA). Move the test lead tip around until a reading is obtained.

6. A resistance reading (typically between 0.995 kΩ to 1.0005 kΩ) should be displayed. The exact reading will vary depending on DMM manufacturer and model. Refer to the user's manual for correct reading. If the DMM reads OL, replace the fuse and test again. Refer to the user's manual or have the DMM serviced if the resistance reading is questionable.

The unit ampere is named after the nineteenth century French physicist André Marie Ampère.

Clamp-On Ammeter Measurement Procedures

Clamp-on ammeters measure current in a circuit by measuring the strength of the magnetic field around a single conductor. Care should be taken to ensure that the meter does not pick up stray magnetic fields. Whenever possible, separate conductors under test from other surrounding conductors by a few inches. If this is not possible, take several readings at different locations along the same conductor.

AC or DC measurement with a clamp-on ammeter or a DMM with a clamp-on current probe accessory follows standard procedures. **See Figure 8-6.**

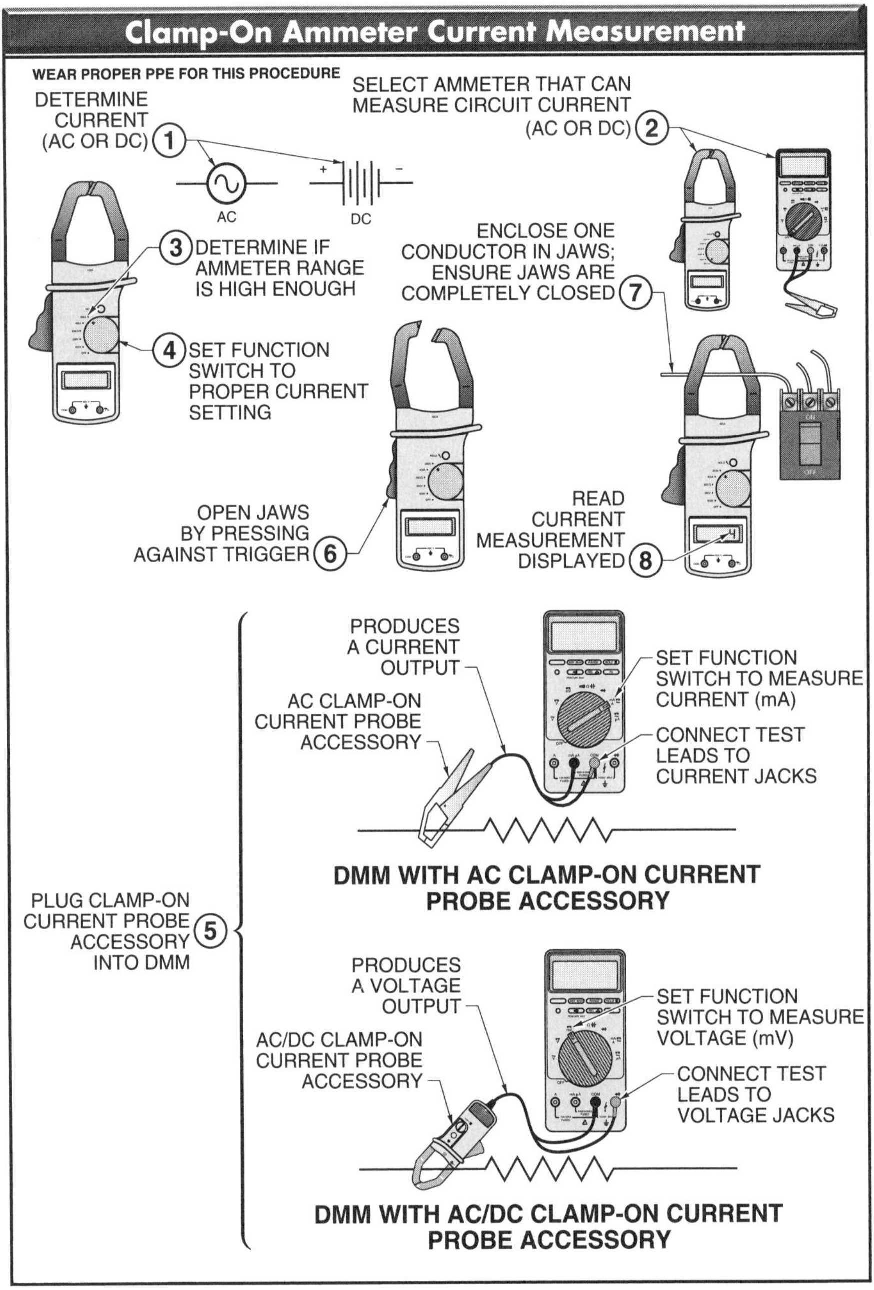

Figure 8-6. Clamp-on ammeters measure current by the strength of the magnetic field around a single conductor.

1. Determine if AC or DC current is to be measured.

2. Select the ammeter required to measure the circuit current (AC or DC). If both AC and DC measurements are required, select an ammeter that can measure both AC and DC.

3. Determine if the ammeter range is high enough to measure the maximum current that may exist in the test circuit. If the ammeter range is not high enough, select an accessory that has a high enough current rating, or select an ammeter with a higher range. If the ammeter includes fused current terminals, check to make sure the ammeter fuses are good.

4. Set the function switch to the proper current setting (600 A, 200 A, 10 A, 400 mA, etc.). Select a setting greater than the highest possible circuit current if there is more than one current position or if the circuit current is unknown.

5. If required, plug the clamp-on current probe accessory into the DMM. The black test lead of the clamp-on current probe accessory is plugged into the common jack. The red test lead is plugged into the mA jack for current measurement accessories that produce a current output. The red test lead is plugged into the voltage (V) jack for current measurement accessories that produce a voltage output. The current measurement accessories that produce current output are designed to measure

AC only and deliver 1 mA to the DMM for every 1 A of measured current (1 mA/A). Current accessories that produce voltage output are designed to deliver 1 mV, 10 mV, or 100 mV to the DMM for every 1 A of measured current (1 mV/A, 10 mV/A, or 100 mV/A).

6. Open the jaws by pressing against the trigger.

7. Enclose one conductor in the jaws. Ensure that the jaws are completely closed before taking readings.

8. Read the current measurement displayed.

Current measurements are taken by enclosing one conductor.

In-Line Current Measurement Procedures

Care is required to protect the ammeter, circuit, and the user when measuring AC or DC current with an in-line ammeter.

WARNING: In-line current measurements should not be taken on high-energy electrical circuits. It is much safer to use a clamp-on ammeter or a DMM with a clamp-on current probe accessory, as the circuit does not have to be opened or interrupted.

Standard safety precautions are followed when using an in-line ammeter. **See Figure 8-7.**

1. Test the DMM fuses following recommended procedures.

2. Ensure that the expected load current measurement is less than the current setting (limit) of the ammeter. Start with the highest current-measuring range if the load current is unknown. If the current measurement may exceed the limit of the in-line ammeter setting, use a clamp-on ammeter, or do not take the measurement.

3. Ensure that the DMM function switch is set to the proper setting for measuring current (AC or DC). Most DMMs include more than one current level (A and mA μA).

4. Ensure that the test leads are connected to the proper jacks for measuring current. Most DMMs include more than one current jack.

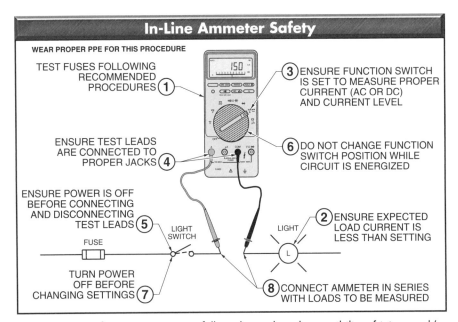

Figure 8-7. Safety precautions are followed to reduce the possibility of injury and/or damage to equipment when taking in-line current measurements.

WARNING: Always ensure that the function switch position matches the connection of the test leads. The DMM can be damaged if the test leads are connected to measure current, and the function switch is set for a different measurement such as voltage or resistance. Some DMMs have Input Alert™ that provides a constant audible warning (beep) if the test leads are connected in the current jacks and a non-current mode is selected.

5. Ensure that the power to the test circuit is OFF before connecting and disconnecting test leads. If necessary, take a voltage measurement to ensure that the voltage is OFF.

6. Do not change the function switch position on the ammeter while the circuit under test is energized.

A flexible current probe accessory allows current measurement without opening the circuit.

7. Turn power to the ammeter and circuit OFF before changing any ammeter settings.

8. Connect the ammeter in series with the load(s) to be measured. Never connect an ammeter in parallel (as with a voltmeter) with the load(s) to be measured.

Many DMMs include a fuse in the current-measuring circuit to prevent damage caused by excessive current. Before using a DMM, check to see if the DMM is fused on the current range being used. The DMM is marked as fused or not fused at the test lead current terminals. In-line current measurements are not recommended if the DMM is not fused. AC or DC is measured with an in-line ammeter using standard procedures. **See Figure 8-8.**

WARNING: Ensure that no body parts contact any part of the live circuit, including metal contact points at the tip of the test leads.

1. Set the function switch to the proper position for measuring the AC or DC current level (A or mA μA). Select a setting high enough to measure the highest possible circuit current if the ammeter has more than one position.

2. Plug the black test lead into the common jack.

3. Plug the red test lead into the current jack. The current jack may be marked A or mA μA.

4. Turn the power to the circuit or device under test OFF and discharge all capacitors if possible.

5. Open the circuit at the test point and connect the test leads to each side of the opening. For DC current, the black (negative) test lead is connected to the negative side of the opening, and the red (positive) test lead is connected to the positive side of the opening. Reverse the black and red test leads if a negative sign appears to the left of the measurement displayed.

6. Turn the power to the circuit under test ON.

7. Read the current measurement displayed.

8. Turn the power OFF and remove the ammeter from the circuit.

CURRENT MEASUREMENT ANALYSIS

Knowing the current measurement in a system, component, circuit, or part of a circuit is very useful when troubleshooting. For example, if the current measurement on a motor is compared with the motor nameplate current rating, the motor-operating condition can be determined. **See Figure 8-9.**

A motor in a car wash system is working to its maximum (fully loaded) if it draws the rated current, which is 8 A in this example. A motor is overloaded if it draws more than the rated current. A motor is underloaded if it draws less than the rated current.

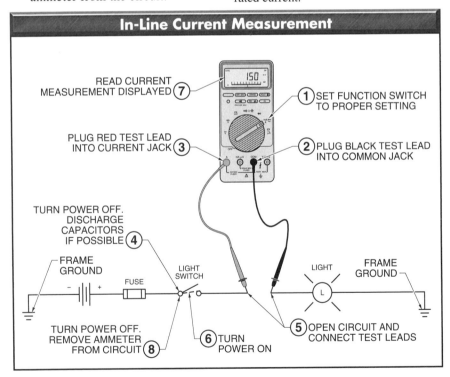

Figure 8-8. An in-line ammeter becomes part of the circuit tested.

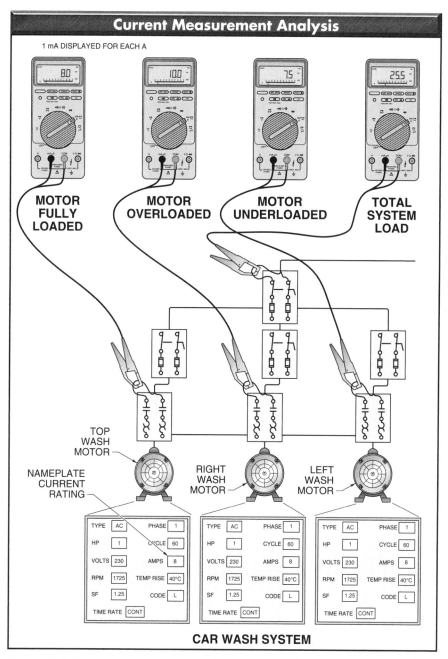

Figure 8-9. Current measurements are taken when testing for motor-operating condition in a system.

The motor size may be increased or the load on the motor decreased if overloads are a problem. Current measurements taken on the power distribution system can be used to indicate total system load.

Generally, higher-than-rated currents usually indicate a problem which can cause additional problems. Higher current produces higher temperatures that cause insulation breakdown and component failure.

Excessive temperature is a major cause of transistor, integrated circuit, and other solid-state electronic component failure. For maximum efficiency, it is recommended that a current measurement be taken when equipment is first installed and during normal operation. These measurements can be used to provide a baseline comparison when troubleshooting a problem in the future.

 One ampere is equal to the flow of approximately 6,250,000,000,000,000,000 electrons per sec past any point in the circuit.

AC voltage measurements are taken when troubleshooting the power supply.

Ohm's Law and Power Formula

OHM'S LAW

Ohm's law is the relationship between voltage (E), current (I), and resistance (R) in a circuit. Ohm's law states that current in a circuit is proportional to the voltage and inversely proportional to the resistance. If the resistance in a circuit remains constant, a change in current is directly proportional to a change in voltage. If voltage in a circuit remains constant, current in a circuit decreases with an increase in resistance, and current in the circuit increases with a decrease in resistance. Using Ohm's law, any value in this relationship can be found when the other two are known. The relationship between voltage, current, and resistance may be visualized by presenting Ohm's law in a pie chart form. **See Figure 9-1.**

Using Ohm's Law

Ohm's law can be used for determining voltage, current, or resistance requirements during circuit design and for predicting circuit characteristics before power is applied. For example, in a heating element (resistive load) circuit, a fixed load resistance of 4 Ω is connected to a variable power supply which supplies 0 V to 24 V. **See Figure 9-2.** The current

in the circuit may be found for any voltage by applying Ohm's law. For example, if the voltage in the circuit is set at 8 V and the resistance is 4 Ω, the current equals 2 A ($8 \div 4 = 2$ A).

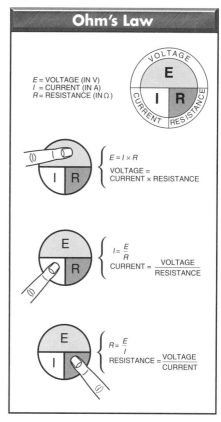

Figure 9-1. Ohm's law is the relationship between voltage (E), current (I), and resistance (R) in a circuit.

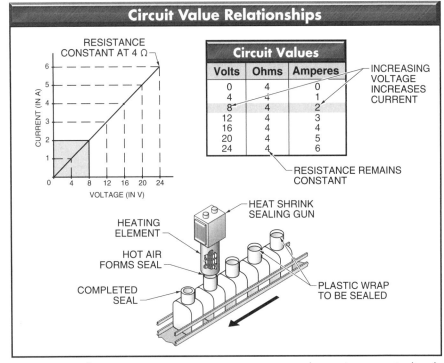

Figure 9-2. If resistance in a circuit remains constant, a change in current is directly proportional to a change in voltage.

In troubleshooting applications, Ohm's law can be used to determine how a circuit should operate and how it is operating under power. For example, when the insulation of equipment or conductors breaks down, a current measurement higher than normal indicates that circuit resistance has decreased or circuit voltage has increased. This information is used when identifying potential problems such as insulation breakdown or a high-voltage condition. In the same circuit, a current measurement lower than normal indicates that the circuit resistance has increased or the circuit voltage has decreased. An increase in circuit resistance is usually caused from poor connections, loose connections, corrosion, and/or damaged components.

When troubleshooting a circuit resistance problem, a method used to determine resistance is to take voltage and current measurements. Resistance cannot be measured in an operating circuit, so the voltage and current measurements are used to determine total circuit resistance by applying Ohm's law. **See Figure 9-3.**

 The Fahrenheit temperature scale was introduced in 1714 by Gabriel Daniel Fahrenheit.

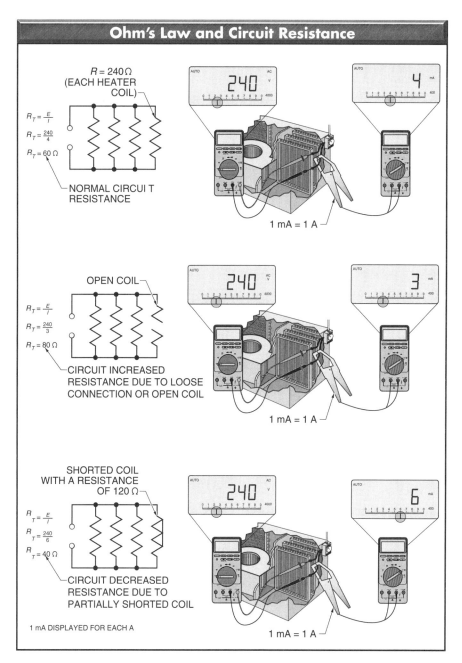

Ohm's Law and Circuit Resistance

$R = 240\,\Omega$
(EACH HEATER COIL)

$R_T = \frac{E}{I}$

$R_T = \frac{240}{4}$

$R_T = 60\,\Omega$

NORMAL CIRCUIT RESISTANCE

1 mA = 1 A

OPEN COIL

$R_T = \frac{E}{I}$

$R_T = \frac{240}{3}$

$R_T = 80\,\Omega$

CIRCUIT INCREASED RESISTANCE DUE TO LOOSE CONNECTION OR OPEN COIL

1 mA = 1 A

SHORTED COIL WITH A RESISTANCE OF 120 Ω

$R_T = \frac{E}{I}$

$R_T = \frac{240}{6}$

$R_T = 40\,\Omega$

CIRCUIT DECREASED RESISTANCE DUE TO PARTIALLY SHORTED COIL

1 mA DISPLAYED FOR EACH A

1 mA = 1 A

Figure 9-3. Ohm's law can be used to determine circuit resistance in operating circuits with voltage and current measurements.

For example, in an electric heater circuit, resistance is determined by measuring circuit voltage and current and applying Ohm's law. Normal circuit resistance is 60 Ω (240 ÷ 4 = 60 Ω). The 60 Ω resistance can help determine the condition of the circuit. For example, if circuit current is 3 A instead of 4 A, circuit resistance has increased from 60 Ω to 80 Ω (240 ÷ 3 = 80 Ω). The 20 Ω gain in resistance could be caused from a loose or dirty connection or an open-coil section.

Open-coil sections increase the total circuit resistance, which decreases circuit current. If circuit current is 6 A instead of 4 A, circuit resistance has decreased from 60 Ω to 40 Ω (240 ÷ 6 = 40 Ω). The 20 Ω loss in resistance could be caused by a partially shorted coil or insulation breakdown.

Cold and Hot Resistance. Because resistance measurements are taken with no operating current in the circuit, the DMM is taking a cold-resistance measurement. *Cold resistance* is the resistance of a component when operating current is not passing through the device. *Hot resistance* is the actual (true) resistance of a component when operating current is passing through the device. Hot resistance is the resistance that determines the amount of current flow. However, hot resistance cannot be measured because DMMs set for measuring resistance cannot be used in an operating circuit.

A cold-resistance measurement in some components is not the same as a hot-resistance measurement. An incandescent lamp, for example, changes resistance once power is applied. The lamp radiates light because current flowing in the circuit raises the filament temperature to a white-hot level. In most materials, resistance increases with an increase in temperature. Good troubleshooting practice requires measuring circuit voltage and current during equipment operation, in addition to measuring resistance before power in the circuit is ON. **See Figure 9-4.**

Loose or dirty connections will increase resistance in the circuit.

Manufacturers of sensitive electronic equipment specify a required maximum ground resistance of 1 Ω or less for proper equipment operation.

Cold and Hot Resistance

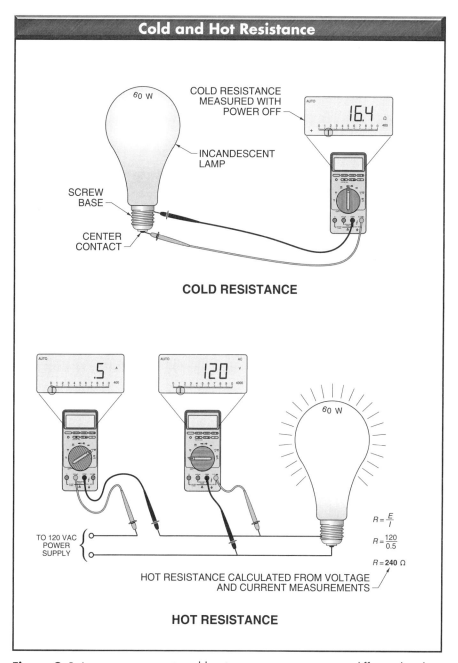

Figure 9-4. In some components, cold-resistance measurements are different than hot-resistance measurements.

POWER FORMULA

The *power formula* is the relationship between power (*P*), voltage (*E*), and current (*I*) in an electrical circuit. Any value in this relationship may be found using the power formula when the other two are known. The relationship between power, current, and voltage may be visualized by presenting the power formula in a pie chart form. **See Figure 9-5.**

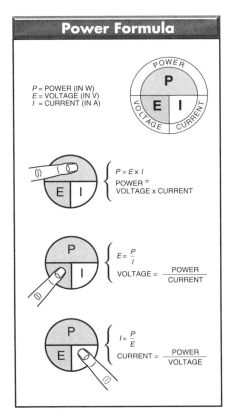

Figure 9-5. The power formula is the relationship between power (*P*), voltage (*E*), and current (*I*) in an electrical circuit.

Using the Power Formula

The power formula is useful for determining expected current values because most electrical equipment lists a voltage and a power rating. The power rating is listed in watts (W) for most appliances and heating elements, or in horsepower (HP) for motors. The power formula is also used to predict circuit characteristics before power is applied. For example, heat is produced any time electricity passes through a wire that has resistance. Heating elements use this principle in devices such as toasters, portable space heaters, hair dryers, and electric water heaters.

When selecting a heating element, the size in watts required is determined by the application. The greater the heat output required, the higher the wattage required. Once the required wattage is determined, the current draw can be calculated using the power formula. **See Figure 9-6.** For example, if a 2000 W heating element is connected to the specified 115 V supply, circuit current is 17.4 A (2000 ÷ 115 = 17.4 A). The circuit current value is then used to determine wire size, switch rating, and circuit breaker rating.

COMBINING OHM'S LAW AND POWER FORMULA

Ohm's law and the power formula may be combined and written as any combination of voltage (*E*), current (*I*), resistance (*R*), or power (*P*). This combination lists six basic formulas and six rearranged formulas. **See Figure 9-7.**

Determining Circuit Current

Heater Element Rating

Catalog Number	Power Rating (W)	Voltage
01	2000	115
02	2000	230
03	3000	115
04	3000	230
05	4000	230
06	5000	230
09	7500	230

$$I = \frac{P}{E}$$

$$I = \frac{2000}{115}$$

$$I = 17.4\ A$$

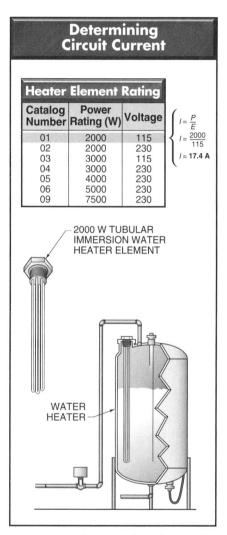

2000 W TUBULAR IMMERSION WATER HEATER ELEMENT

WATER HEATER

Figure 9-6. The power formula is used to predict circuit operating characteristics for sizing wire, switches, and circuit breakers.

In the National Electrical Code®, the term watts (W) has been generally superseded by the term volt-amperes (VA) for computation of loads. References to nameplate ratings still reflect the term watts on certain loads.

Ohm's Law and Power Formula

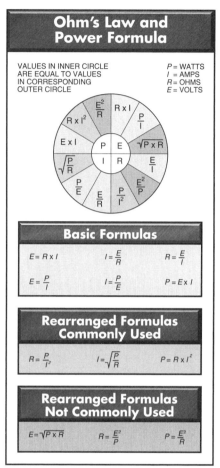

VALUES IN INNER CIRCLE ARE EQUAL TO VALUES IN CORRESPONDING OUTER CIRCLE

P = WATTS
I = AMPS
R = OHMS
E = VOLTS

Basic Formulas

$$E = R \times I \qquad I = \frac{E}{R} \qquad R = \frac{E}{I}$$

$$E = \frac{P}{I} \qquad I = \frac{P}{E} \qquad P = E \times I$$

Rearranged Formulas Commonly Used

$$R = \frac{P}{I^2} \qquad I = \sqrt{\frac{P}{R}} \qquad P = R \times I^2$$

Rearranged Formulas Not Commonly Used

$$E = \sqrt{P \times R} \qquad R = \frac{E^2}{P} \qquad P = \frac{E^2}{R}$$

Figure 9-7. Ohm's law and the power formula may be combined to obtain additional rearranged formulas.

OHM'S LAW AND IMPEDANCE

Ohm's law and the power formula are limited to circuits in which electrical resistance is the only significant opposition to the flow of current. This limitation includes all direct current (DC) circuits and any alternating current (AC) circuits that do not contain a significant amount of inductance and/or capacitance.

Inductance (L) is the property of a circuit that causes it to oppose a change in current due to energy stored in a magnetic field. *Capacitance (C)* is the ability of a component or circuit to store energy in the form of an electrical charge. Capacitance produces an opposition to the flow of current in a circuit when connected to an AC power supply.

AC circuits not having inductance and/or capacitance include components such as heating elements and incandescent lamps. AC circuits having inductance include a coil as the load. All motors, transformers, and solenoids have a coil. AC circuits having capacitance include a capacitor such as a capacitor start-and-run motor.

In DC circuits and AC circuits that do not contain a significant amount of inductance and/or capacitance, the opposition to the flow of current is resistance *(R)*. In circuits that contain inductance or capacitance, the opposition to the flow of current is reactance *(X)*. Inductive reactance is indicated by X_L. Capacitive reactance is indicated by X_C.

In circuits that contain resistance *(R)* and reactance *(X)*, the combined opposition to the flow of current is impedance *(Z)*. *Impedance* is the total opposition of any combination of resistance, inductive reactance, and capacitive reactance offered to the flow of alternating current. Impedance is expressed in ohms. **See Figure 9-8.**

Ohm's law is used in circuits that contain impedance, but *Z* is substituted for *R* in the formula. The letter *Z* represents the total resistive force (resistance and reactance) opposing current flow. The relationship between voltage *(E)*, current *(I)*, and impedance *(Z)* may be visualized by presenting the relationship in pie chart form.

A clamp-on ammeter measures current by the strength of the magnetic field produced around a single conductor.

 A noncontact infrared temperature probe can be used for taking temperature measurements on operating motors, bearings, electrical panels, moving objects, or hazardous materials.

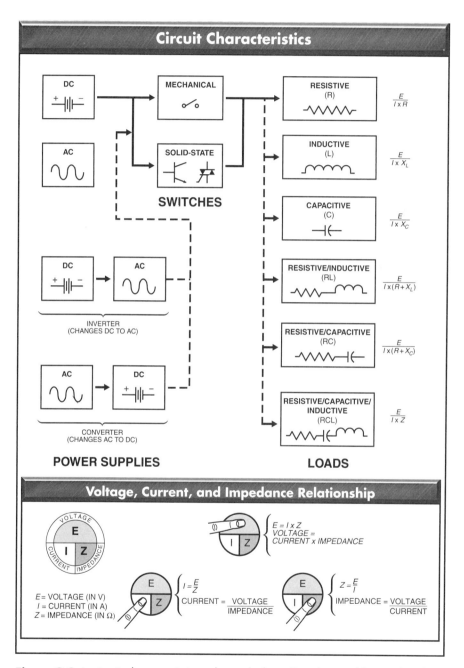

Figure 9-8. In circuits that contain impedance, the letter *Z* is substituted for *R* in the Ohm's law formula.

Measuring Frequency and Duty Cycle

FREQUENCY

Frequency is the number of cycles per second (cps) in an AC sine wave. Frequency is measured in hertz (Hz). A *hertz (Hz)* is the international unit of frequency equal to 1 cycle per second. A *cycle* is one complete wave of alternating voltage or current. An *alternation* is one half of a cycle. A *period* is the time required to produce one complete cycle of a waveform. **See Figure 10-1.**

The term frequency is typically used to describe electrical equipment operation, as when discussing power line frequency (which is normally 50 Hz or 60 Hz) and variable frequency drives (which normally use a 1 kHz to 20 kHz carrier frequency). The term frequency is also used to describe electronic equipment operation, as in the terms audio frequency (15 Hz to 20 kHz) and radio frequency (30 kHz to 300 kHz for low frequency, 300 kHz to 3 MHz for medium frequency, 3 MHz to 30 MHz for high frequency, and 30 MHz to 300 MHz for very high frequency).

Frequency Measurement

Circuits and equipment may be designed to operate at a fixed or variable frequency. Equipment designed to operate at a fixed frequency performs abnormally if operated at a different frequency than specified. For example, an AC motor designed to operate at 60 Hz operates slower if the frequency is less than 60 Hz and operates faster if the frequency is more than 60 Hz. For AC motors, any change in frequency causes a proportional change in motor speed. For example, a 5% reduction in frequency produces a 5% reduction in motor speed.

A DMM that includes a Frequency Counter mode can be used to measure frequency when troubleshooting electrical and electronic equipment. *Frequency Counter mode* is a DMM mode that measures the frequency of AC signals. If the DMM also includes a MIN MAX Recording mode, frequency measurements can be recorded over a specific time period. The MIN MAX Recording mode is used to record frequency measurements the same way voltage, current, or resistance measurements are recorded. A DMM that includes an Autorange mode automatically selects the frequency measurement range. However, if the frequency of the measured voltage is outside the frequency measurement range, the DMM cannot display an accurate measurement. Refer to the user's manual for specific frequency measurement ranges.

 Current flow can be used to produce effects such as light, heat, or magnetic fields.

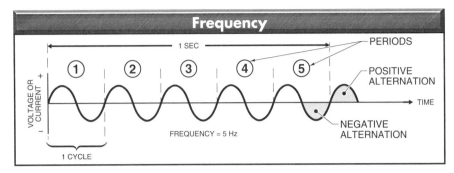

Figure 10-1. Frequency is the number of cycles produced per second measured in hertz (Hz).

Frequency Measurement Procedures

Frequency measurement using a DMM with a Hz setting follows standard procedures. **See Figure 10-2.**

1. Set the function switch to Hz.

2. Plug the black test lead into the common jack.

3. Plug the red test lead into the voltage jack.

4. Connect the black test lead first and red test lead second.

5. Read the frequency measurement displayed. The abbreviation Hz should be displayed to the right of the frequency measurement.

Frequency measurement using a DMM with a Hz button follows standard procedures. **See Figure 10-3.**

1. Set the function switch to AC voltage. Set the range to the highest voltage setting if the voltage in the circuit is unknown. Most DMMs power up in the Autorange mode, which automatically selects the measurement range based on the voltage present.

2. Plug the black test lead into the common jack.

3. Plug the red test lead into the voltage jack.

4. Connect the test leads to the circuit. The position of the test leads is arbitrary.

5. Read the voltage displayed.

6. With the DMM still connected to the circuit, press the Hz button.

7. Read the frequency measurement displayed. The abbreviation Hz should be displayed to the right of the frequency measurement.

In some circuits, there may be enough distortion on the line to prevent an accurate frequency measurement. For example, AC variable frequency drives (VFDs) can produce frequency distortions. When testing VFDs, use the 🔲 VAC setting for accurate readings. To accurately check the frequency on meters without the 🔲 VAC setting, the function switch can be moved from AC voltage To accurately check the LOVAC setting, the function switch can be moved from AC to DC voltage and the Hz button pressed again to measure the frequency on the DC voltage setting. The DMM can be changed to DC voltage because it is still set to measure frequency rather than voltage. If there is no distortion affecting the frequency measurement, the displayed frequency measurement is the same for AC voltage and DC voltage. If two different frequency measurements are displayed, the lower frequency measurement is usually the correct measurement.

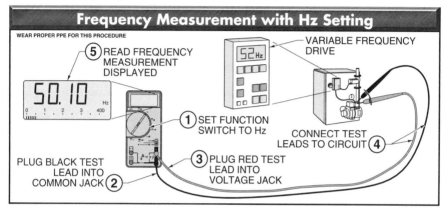

Figure 10-2. Frequency measurement is displayed after connecting test leads to the circuit when using a DMM with a Hz setting.

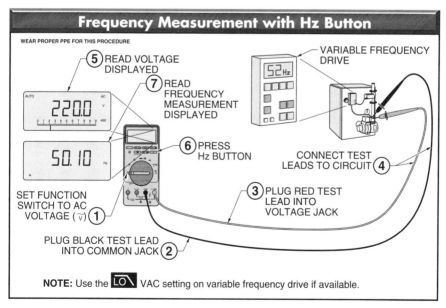

Figure 10-3. Frequency measurement using a DMM with a Hz button requires first measuring AC voltage.

Motor Drive Frequency Measurement —Safe Work Practices

A maintenance service call requires reprogramming the maximum operating frequency parameter of a motor drive from 60 Hz to 50 Hz to ensure that the machine operator cannot operate the machine too fast. After the motor drive parameter is reprogrammed, a frequency measurement will be taken at the motor connection points on the drive (T1/U, T2/V, T3/W). The motor drive's input and output voltages are 230 VAC and powered from a motor control center (MCC) located in a different area.

MINIMUM RECOMMENDATIONS AT MEASURING POINTS:
CAT III Rated DMM and Test Leads
Hazard/Risk Category #1

REQUIRED PPE
* protective long-sleeve shirt rated 4 cal/cm^2 and pants or coveralls rated 4 cal/cm^2 (or untreated cotton denim jeans)
* safety glasses
* Class 00 (500 V) rubber insulating gloves with leather protectors
* hard hat

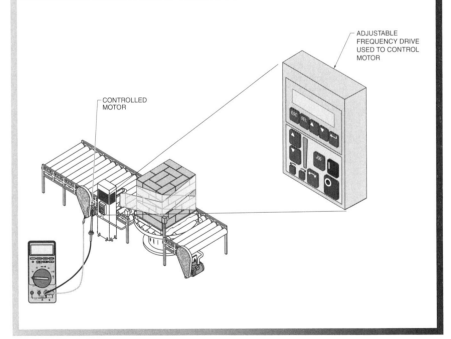

Measuring Frequency with Clamp-On Meter

Some clamp-on meters have the ability to measure frequency in the same simple way they measure current. **See Figure 10-4.** To measure frequency with a clamp-on meter, apply the following procedure:

CAUTION: The DMM can be changed from AC voltage to DC voltage when in the Frequency Counter mode, but the DMM should never be taken out of Frequency Counter mode until the test leads are removed from the circuit.

1. Determine if the clamp-on meter range is high enough to measure the maximum Hertz that may exist in the test circuit. If the clamp-on meter range is not high enough, select a clamp-on meter with a higher range.

2. Access the wire to be tested.

3. Set the function switch to frequency.

4. Open the jaws by pressing against the trigger.

5. Enclose one conductor in the jaws. Ensure that the jaws are completely closed and the conductor is positioned at the intersection of the alignment marks before taking the reading.

6. Read the frequency measurement displayed on the clamp-on meter.

7. Remove the clamp-on meter from the wire.

8. Turn the clamp-on meter OFF to prevent battery drain. *Note:* A DMM clamp-on current probe accessory can also be connected to any DMM that includes a Hz function to measure frequency.

DUTY CYCLE

Duty cycle (duty factor) is an alternative frequency measurement that is the ratio of time a load or circuit is ON to the time a load or circuit is OFF, expressed as a percentage. *Duty Cycle mode* is a DMM mode that measures the duty cycle of a load or circuit. Many loads are rapidly turned ON and OFF by a fast-acting electronic switch to accurately control output power at the load. For example, load operation such as lamp brightness, heating element outputs, and magnetic strength of a coil can be controlled by ON and OFF time periods or cycles per second. Any load that is turned ON and OFF several times per second has a duty cycle. **See Figure 10-5.**

Duty cycle is expressed as a percentage and is determined by the amount of ON time divided by the total time of the cycle multiplied by 100. To find duty cycle, apply the formula:

$$DC = \frac{OT}{T} \times 100$$

where

DC = duty cycle

(in percent of ON time)

OT = ON time (in sec)

T = total time (in sec)

For example, in an automotive electronic fuel injection system, voltage pulses supplied to the fuel injector valve solenoid control the fuel injector valve at a fixed rate of 10 cps. If a fuel injector valve is pulsed ON at variable durations (pulse-width modulation), the duty cycle varies. **See Figure 10-6.** What is the duty cycle if the fuel injector valve is pulsed ON for 0.01 sec out of a 0.1 sec cycle?

$$DC = \frac{OT}{T} \times 100$$

$$DC = \frac{0.01}{0.1} \times 100$$

$$DC = \mathbf{10\%}$$

If the fuel injector valve is pulsed ON for 0.05 sec out of a 0.1 sec cycle, the fuel injector duty cycle equals 50%. If the fuel injector valve is pulsed ON for 0.09 sec out of a 0.1 sec cycle, the fuel injector duty cycle equals 90%.

Pulse-width modulation allows fuel supplied to the engine to be precisely controlled electronically. The voltage average for each duty cycle is determined by the amount of pulse ON time.

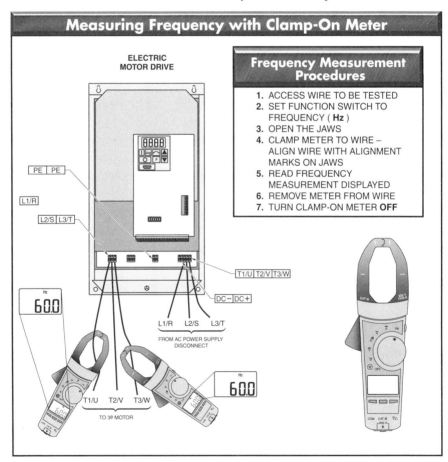

Measuring Frequency with Clamp-On Meter

ELECTRIC
MOTOR DRIVE

Frequency Measurement Procedures

1. ACCESS WIRE TO BE TESTED
2. SET FUNCTION SWITCH TO FREQUENCY (**Hz**)
3. OPEN THE JAWS
4. CLAMP METER TO WIRE – ALIGN WIRE WITH ALIGNMENT MARKS ON JAWS
5. READ FREQUENCY MEASUREMENT DISPLAYED
6. REMOVE METER FROM WIRE
7. TURN CLAMP-ON METER **OFF**

PE PE
L1/R
L2/S L3/T

T1/U T2/V T3/W
DC– DC+

L1/R L2/S L3/T
FROM AC POWER SUPPLY DISCONNECT

Hz
60.0

Hz
60.0

T1/U T2/V T3/W
TO 3∅ MOTOR

Figure 10-4. The frequency measurement of an electric motor drive output current can be used to calculate the speed of a motor.

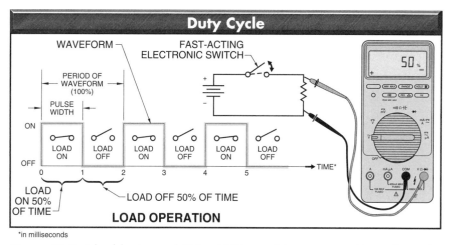

Figure 10-5. A load that is turned ON and OFF several times per second has a duty cycle.

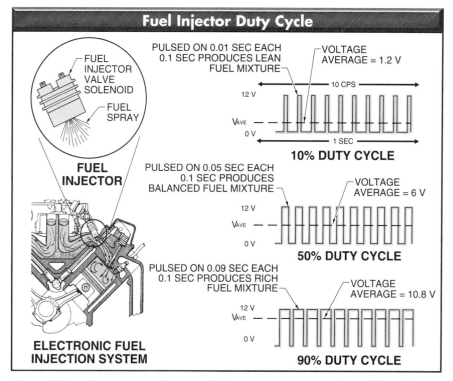

Figure 10-6. An automotive electronic fuel injection system uses pulse-width modulation for precise control of fuel supplied to the engine.

To complete a cycle, the power must go from OFF to ON to OFF again. When measuring duty cycle, the DMM displays the amount of time the input signal is above or below a fixed trigger level. **See Figure 10-7.** *Trigger level* is the fixed level at which the DMM counter is triggered to record frequency. *Slope* is the waveform edge on which the trigger level is selected.

The percent of time above the trigger level is displayed if the positive trigger slope is selected. The percent of time below the trigger level is displayed if the negative trigger slope is selected. The slope selected is indicated by a positive (+) or negative (–) symbol on the DMM display. Most DMMs default to display the positive trigger slope. The negative trigger slope is usually selected by pressing an additional button. Refer to the user's manual for specific instructions.

 DMM replacement fuses must match manufacturer's specifications to prevent instrument damage and/or personal injury.

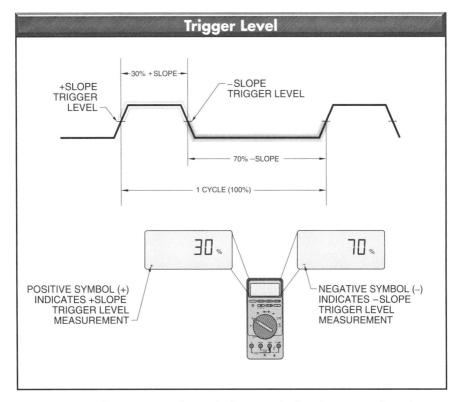

Figure 10-7. When measuring duty cycle, the DMM displays the amount of time the input signal is above or below a fixed trigger level.

Duty Cycle Measurement Procedures

Duty cycle measurement with a DMM follows standard procedures. **See Figure 10-8.**

1. Set the DMM to measure duty cycle. On most DMMs, the Hz button is pressed twice and the DMM is ready to measure the duty cycle when a percent sign (%) appears on the right side of the display.

2. Plug the black test lead into the common jack.

3. Plug the red test lead into the voltage jack.

4. Connect the test leads to the circuit. Connect the black test lead to the circuit ground (negative polarity test point) and the red test lead to the positive test point. Reverse the black test lead and red test lead if a negative symbol (–) appears in display. In AC circuits, the position of the test leads is arbitrary.

5. Read the duty cycle measurement displayed. A positive symbol (+) indicates ON time percent voltage measurement. A negative symbol (–) indicates OFF time percent voltage measurement.

6. Press the beeper button to toggle between ON time percent voltage measurement and OFF time percent voltage measurement. The button used varies with the DMM. Refer to the user's manual for specific instructions.

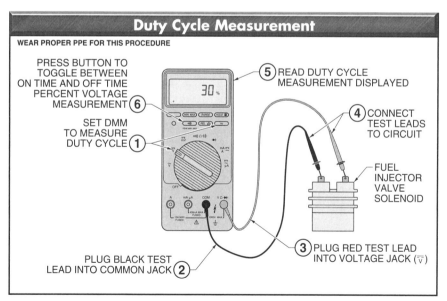

Figure 10-8. On most DMMs, the Hz button is pressed twice for Duty Cycle mode.

Frequency measurements are taken when troubleshooting variable frequency drives.

Testing Diodes

DIODES

A *diode* is an electronic device that allows current to flow in only one direction. Diodes are also called rectifiers because they change (rectify) AC into pulsating DC. Diodes are rated according to their function in the circuit, voltage, and current capacity. A diode has polarity as determined by an anode and a cathode. An *anode* is the positive lead of a diode. A *cathode* is the negative lead of a diode. Diodes are available in different configurations for specific applications. **See Figure 11-1.**

Voltage measurements are used when troubleshooting electronic circuit boards.

Forward bias is the condition of a diode when a diode allows current flow. The anode in a forward-biased diode has a positive polarity compared to the cathode. *Reverse bias* is the condition of a diode when it does not allow current flow and acts as an insulator. The cathode of a reverse-biased diode has a positive polarity compared to the anode.

A diode allows current to flow only when negative voltage is applied to the cathode (forward bias). The diode acts as a closed switch when it is forward-biased and as an open switch when it is reverse-biased. A diode is rated for the maximum forward-bias current it can safely conduct and the maximum reverse-bias voltage that can be applied. **See Figure 11-2.**

The maximum forward-bias current rating applies to the amount of current the diode can withstand while conducting electrons in the forward-bias direction. Exceeding the maximum forward-bias current rating causes the diode to overheat, resulting in possible diode failure. A maximum reverse-bias voltage rating applies to the amount of voltage that may be applied in the reverse direction without allowing current flow in the opposite direction (voltage breakover).

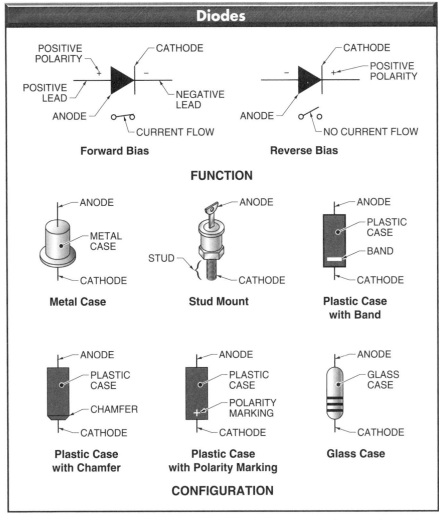

Figure 11-1. A diode is an electronic device that allows current to flow in only one direction and is available in different configurations.

If the diode rating is exceeded, the diode can fail. Diode failure occurs as the diode shorts and allows current flow in both directions. Failure also occurs when the diode opens and does not allow current flow in either direction.

Proper diode function can be tested with a DMM.

 In 1977, the Fluke 8020 Series DMM was introduced as the first true hand-held DMM.

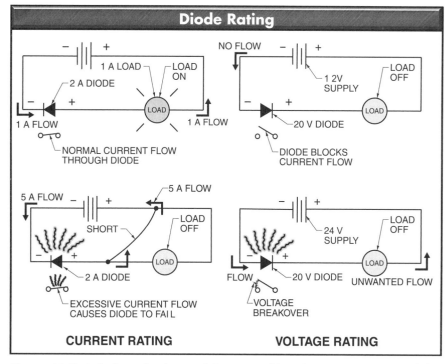

Figure 11-2. A diode can be damaged if current and/or voltage ratings are exceeded.

DIODE TEST PROCEDURES

Diodes can be tested using the DMM Diode Test mode or the DMM Resistance mode. The Diode Test mode is the first choice and best method for diode testing because a resistance measurement does not always indicate whether a diode is good or bad. In addition, a resistance measurement should not be taken when a diode is connected in a circuit as it produces a false reading. However, once a diode is determined bad by using the Diode Test mode, a resistance measurement can be used to verify the diode is bad in the specific application. The DMM Resistance mode is also used if the DMM does not include a Diode Test mode.

Of all switching circuits, 98% are AC.

Diode Test with Diode Test Mode Procedures

A diode is best tested by measuring the voltage drop across the diode when it is forward-biased. A good diode has a voltage drop across it when it is forward-biased and is allowing current to flow. The DMM Diode Test mode produces a small voltage between the test leads. The DMM then displays the voltage drop when the test leads are connected across a diode. Diode testing using the

DMM Diode Test mode follows standard procedures. **See Figure 11-3.**

1. Ensure that all power to the circuit is OFF and there is no voltage at the diode. Voltage may be present in the circuit from charged capacitors. Set the DMM to measure AC or DC voltage as required.

2. Set the DMM function switch to Diode Test mode.

3. Connect the test leads to the diode. Record the measurement displayed.

4. Reverse the test leads. Record the measurement displayed.

A good forward-biased diode displays a voltage drop ranging from 0.5 V to 0.8 V for the most commonly used silicon diodes. Some germanium diodes have a voltage drop ranging from 0.2 V to 0.3 V. The DMM displays OL when a good diode is reverse-biased. The OL reading indicates the diode is functioning as an open switch.

A bad (opened) diode does not allow current to flow in either direction. The DMM then displays OL in both directions when the diode is opened. A shorted diode has the same voltage drop reading (approximately 0.4 V) in both directions.

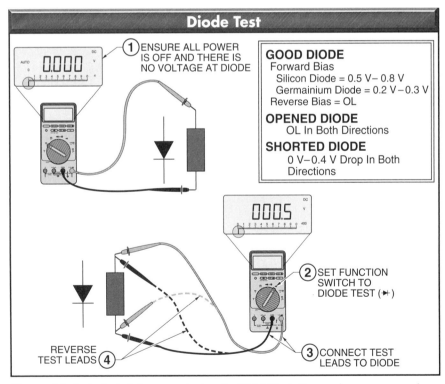

Figure 11-3. A diode is best tested using the DMM Diode Test mode to measure voltage drop across the diode when it is forward-biased.

Diode Testing—Safe Work Practices

A maintenance service call states that the cylinder-operated pistons used to push bread pans into an oven are sticking and not returning to their normal position. The machine operators in the area also state that the control panel below the cylinders seems to be making more noise when the pistons stick. It is known that if the diode under the solenoid-operated valve is short-circuited, AC will be applied to the solenoid and can cause this type of problem. The entire electrical circuit operates on 24 VAC in the control cabinet with diodes used to rectify AC to DC as needed.

MINIMUM RECOMMENDATIONS AT MEASURING POINTS:
CAT II Rated DMM and Test Leads
Hazard/Risk Category #0

REQUIRED PPE
- protective long-sleeve denim or cotton shirt and pants or coveralls
- safety glasses
- Class 00 (500 V) rubber insulating gloves with leather protectors
- hard hat

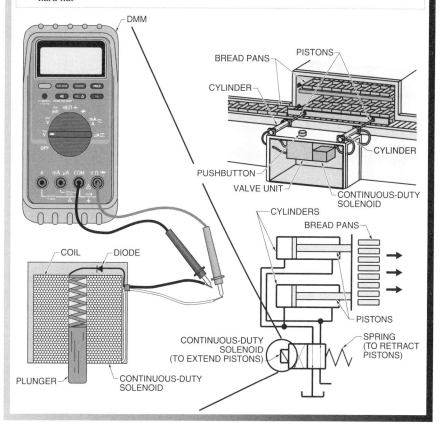

Diode Test with Resistance Mode Procedures

A DMM set to Resistance mode is used as an additional diode test, or when the DMM does not include the Diode Test mode. **See Figure 11-4.** The diode is forward-biased when the positive (red) test lead is on the anode and the negative (black) test lead is on the cathode. The forward-biased resistance of a good diode should range from 1000 Ω to 10 MΩ. The resistance measurement is high when the diode is forward-biased because current from the DMM voltage source flows through the diode. This causes the high-resistance measurement required for testing.

The diode is reverse-biased when the positive (red) test lead is on the cathode and the negative (black) test lead is on the anode.

The reverse-biased resistance of a good diode displays OL on a DMM. The diode is bad if readings are the same in both directions.

Diode testing using the DMM Resistance mode follows standard procedures.

1. Ensure that all power to the circuit is OFF and there is no voltage at the diode. Voltage may be present in the circuit from charged capacitors. Set the DMM to measure AC or DC voltage as required.
2. Set the function switch to Resistance mode.
3. Connect the test leads to the diode removed from the circuit. Record the measurement displayed.
4. Reverse the test leads. Record the measurement displayed.
5. When using the Resistance mode to test diodes, compare the readings taken with a known good diode for best results.

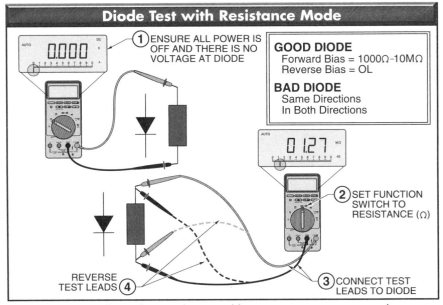

Figure 11-4. Diode condition is determined by resistance measurements when testing with the DMM Resistance mode.

Measuring Capacitance

DIGITALMULTIMETERPRINCIPLES

CAPACITANCE

Capacitance is the ability of a component or circuit to store energy in the form of an electrical charge. A *capacitor* is an electrical device designed to store electrical energy by means of an electrostatic field. Capacitors are also known as condensers in the automotive, marine, and aviation fields.

A capacitor consists of two conductors (plates) separated by an insulator (dielectric). The dielectric allows an electrostatic charge to develop. The electrostatic charge becomes a source of stored electrical energy. The strength of the charge is determined by the applied voltage, size of the conductors, and quality of the insulation. The dielectric may be air, paper, oil, ceramic, mica, or other nonconducting material. **See Figure 12-1.** Capacitors are commonly used as filters in AC circuits, to block DC voltages in electronic circuits, to prevent unwanted feedback problems, and to improve torque in capacitor start-and-run AC motors.

The measurement unit of capacitance is the farad (F). The farad is a unit that is too large for most electrical/electronic applications. Capacitance and capacitor values are usually expressed and displayed in smaller units such as microfarads (μF) and nanofarads (nF). A microfarad (μF) is

equal to 0.000001 F. A nanofarad (nF) is equal to 0.000000001 F, and 1 μF is equal to 1000 nF. Capacitors have a limited life and are often the cause of a malfunction. Capacitors may have a short circuit, an open circuit, and/or may physically deteriorate to the point of failure. When a capacitor short circuits, a fuse may blow and/or other circuit components may be damaged. When a capacitor opens or deteriorates, the circuit and/or circuit components may not operate. Deterioration can also change the capacitance value of a capacitor, causing additional problems.

Many DMMs now include direct capacitance measurements to 100,000 μF. Capacitors can be tested using the Capacitance Measurement mode or Resistance mode on a DMM. The Capacitance Measurement mode directly measures and displays capacitor values up to several thousand μF on most DMMs. When a DMM does not have a Capacitance Measurement mode, or the value of the capacitor is higher than the rated capacitance measuring range of the DMM (see user's manual for range) the Resistance mode can be used to test the capacitor.

In 1917, a group of engineers met to introduce order in industrial work by creating standards. This led to the formation of the Canadian Standards Association (CSA).

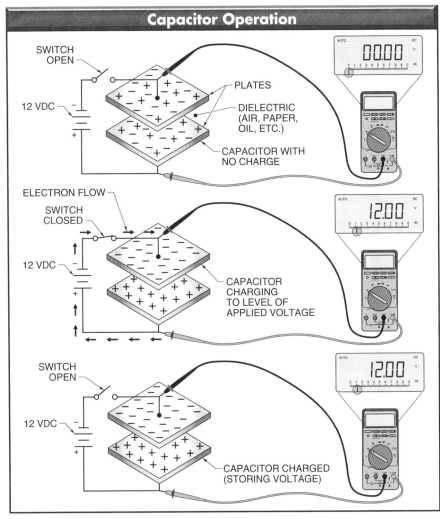

Figure 12-1. A capacitor stores energy in the form of an electrical charge.

Measuring Capacitance with Capacitance Measurement Mode Procedures

The DMM Capacitance Measurement mode directly measures and displays lower capacitive values. Capacitance is measured by the DMM charging the capacitor with a known current, measuring the resultant voltage, and calculating the capacitance.

 In 1977, the Model 8000A DMM was the first Fluke product to use a light-emitting diode (LED) display.

WARNING: A good capacitor stores an electrical charge and may be energized when power is removed. Before touching a capacitor or taking a capacitance measurement, turn all power OFF and discharge the capacitor. Use a DMM to ensure power is OFF. To safely discharge a capacitor after power is removed, connect a 20,000 Ω (20 kΩ), 5 W resistor across the capacitor terminals for 5 sec. On very large capacitors, the process may have to be repeated as the capacitor can build up a charge again after discharging. Use a DMM to confirm the capacitor is fully discharged.

Capacitance measurement using the DMM Capacitance Measurement

mode follows standard procedures. **See Figure 12-2.**

Electric motor horsepower ratings range from fractional HP to nearly 10,000 HP.

1. Ensure that all power to the circuit is OFF by testing with a DMM. Set the DMM to measure AC voltage if the capacitor is used in an AC circuit or DC voltage if the capacitor is used in a DC circuit.

2. Visually inspect the capacitor for leakage, cracks, bulges, or other deterioration. Replace the capacitor if signs of deterioration are present.

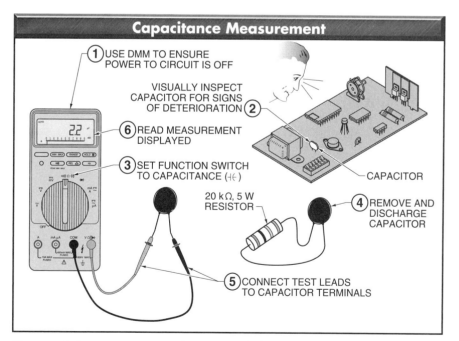

Capacitance Measurement

① USE DMM TO ENSURE POWER TO CIRCUIT IS OFF

VISUALLY INSPECT CAPACITOR FOR SIGNS OF DETERIORATION ②

⑥ READ MEASUREMENT DISPLAYED

③ SET FUNCTION SWITCH TO CAPACITANCE (⊣⊢)

CAPACITOR

20 kΩ, 5 W RESISTOR

④ REMOVE AND DISCHARGE CAPACITOR

⑤ CONNECT TEST LEADS TO CAPACITOR TERMINALS

Figure 12-2. A capacitor may be energized when power is removed and must be discharged before touching or taking a measurement.

3. Set the function switch to Capacitance Measurement mode. The function switch must be moved to Capacitance Measurement mode on most DMMs, and a function change button must be pressed. See user's manual.

4. Remove the capacitor from the circuit. Discharge the capacitor by connecting a 20,000 Ω, 5 W resistor across the capacitor terminals for 5 sec. When measuring low capacitance values, the Relative Mode can be used to remove the capacitance of the test leads. *Note:* To place the DMM in the Relative mode, touch the meter leads together and press the REL function button.

5. Connect the test leads to the capacitor terminals. Keep test leads connected for a few seconds to allow the DMM to automatically select the proper range.

6. Read the measurement displayed. If the capacitance value is within the measurement range, the DMM displays the value of the capacitor. The DMM displays OL if the capacitance value is higher than the measurement range, or if the capacitor is faulty.

Check the capacitor for leakage, cracks, bulges, or other deterioration if the capacitance is outside the measurement range. Replace the capacitor if deterioration is present, and test using the DMM Resistance mode. The DMM Capacitance Measurement mode is useful when identifying the most obvious capacitor failures. In some cases, the capacitor may test good but not function properly.

Testing Capacitor with Resistance Mode Procedures

The DMM Resistance mode is used to test capacitors that have a larger capacitance than the DMM Capacitance Measurement mode range. In addition, the Resistance mode can be used to test a capacitor for its ability to be charged and hold a charge.

To test a capacitor for its ability to be charged, a DMM is set to measure capacitor resistance over time. A capacitor that does not have a charge (fully discharged) starts with a low-resistance value when first connected to the DMM. Resistance starts to increase as the capacitor is charged. The charging process is indicated on a DMM by a consistent increase in resistance measurement until the DMM displays OL. A shorted capacitor does not charge and the DMM displays a constant zero or a low-resistance reading. A shorted capacitor is bad and must be replaced.

To test a capacitor for its ability to be charged, the DMM is used to charge the capacitor to a set resistance point. At the set resistance point, the test leads are removed. The DMM reading should continue at approximately the same resistance measurement, and resistance should continue to increase when the test leads are reconnected. A capacitor that cannot hold a charge is bad and must be replaced. Capacitor testing using the DMM Resistance mode follows standard procedures. **See Figure 12-3.**

The International Organization for Standardization (ISO) is a nongovernmental international organization comprised of national standards institutions from over 90 countries (one per country).

Capacitor Test with Resistance Mode

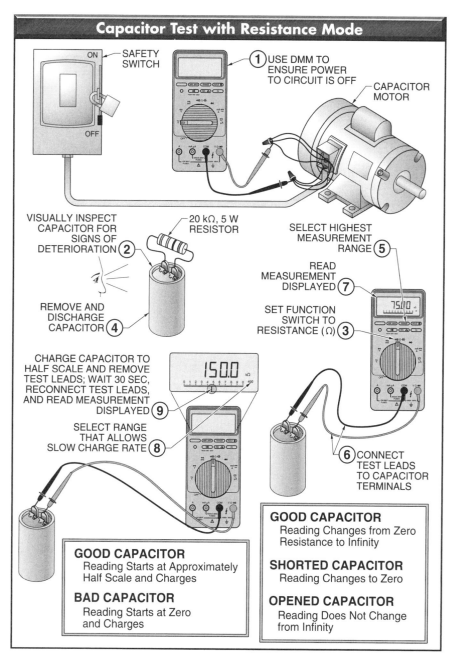

Figure 12-3. The DMM Resistance mode is used to test capacitors that have a larger capacitance than the DMM Capacitance Measurement mode range.

1. Ensure that all power to the circuit is OFF by testing with a DMM. Set the DMM to measure AC voltage if the capacitor is used in an AC circuit or DC voltage if the capacitor is used in a DC circuit.

WARNING: A good capacitor stores an electrical charge and may be energized when power is removed. Before touching a capacitor or taking a capacitance measurement, turn all power OFF and discharge the capacitor. Use a DMM to ensure power is OFF. To safely discharge a capacitor after power is removed, connect a 20,000 Ω, 5 W resistor across the capacitor terminals for 5 sec. Use a DMM to confirm the capacitor is fully discharged.

2. Visually inspect the capacitor for leakage, cracks, bulges, or other deterioration. Replace the capacitor if signs of deterioration are present.

3. Set the function switch to Resistance mode.

4. Remove the capacitor from the circuit and discharge it. To safely discharge a capacitor, connect a 20,000 Ω, 5 W resistor across the terminals for 5 sec.

5. If the DMM has a Manual Range mode, set the DMM to the highest resistance measurement range. If the DMM is in Autorange mode, the DMM automatically selects the best range setting. The best range setting varies with the capacitor tested. Several range settings may have to be tried in order to determine the best setting for the specific capacitor.

6. After the capacitor is discharged, connect the test leads to the capacitor terminals. The resistance reading of the capacitor should be displayed and continuously increase in value, until the DMM displays OL. The amount of time the DMM takes to reach an overload reading depends on the range setting. A range setting that allows the resistance to increase over a 10 sec or more time period is preferred. Several range settings can be tested. Before testing a different range setting, discharge the capacitor.

The capacitor is charged only by the low DC power supply (battery) in the DMM, and a low-resistance conductor can be used to discharge the capacitor before each additional test. If the DMM includes a bar graph, capacitor performance can be monitored by changes in the bar graph.

7. If the resistance reading of the DMM remains low and the capacitor does not charge, the capacitor is bad (shorted) and must be replaced. If the resistance reading of the DMM remains very high (OL), the capacitor is bad (opened) and must be replaced. If the resistance reading indicates that the capacitor is charging, then it must be tested for its ability to hold a charge.

8. To test the ability of a capacitor to hold a charge, select a resistance measurement range that allows the capacitor to charge for 10 sec or more.

9. With the capacitor fully discharged, connect the test leads to the capacitor, and allow the DMM to charge to approximately half the resistance point. If the DMM includes a bar graph, allow the capacitor to charge until the pointer is at approximately half scale. Note the DMM reading at this point and remove the test leads. Wait approximately 30 sec and reconnect the test leads. The measurement should start at approximately the same point as when the test leads were removed if the capacitor can hold a charge. If the resistance measurement starts back at the low-resistance starting point, the capacitor cannot hold a charge and must be replaced. When reconnecting the test leads, make sure the positive (red) and the negative (black) test leads are connected to the same capacitor terminals. If the test leads are reversed, the capacitor discharges, indicated as a decreasing resistance on the display.

DMM ACCESSORY APPLICATIONS

DMMs are commonly designed to take electrical measurements such as AC voltage, DC voltage, and resistance. Accessories are used with DMMs for specific tasks. A *DMM accessory* is a separate connecting and measuring device used with a DMM to enhance usefulness and efficiency. The specific accessory required depends on the intended use. A DMM used for a variety of troubleshooting applications should have the proper accessories available. A holster should be used if the DMM is used in a harsh environment, or requires standing or hanging when taking measurements. A carrying case is used to store accessories and protect the DMM when not in use. Some DMMs are designed to accept a magnetic hanging accessory, which makes the DMM easier to use for certain measurements.

TEST LEADS, TEST CLIPS, AND TEST PROBES

DMMs shipped from the manufacturer commonly include a pair of test leads. A *test lead* is a flexible, insulated lead that serves as a conductor from the accessory to the DMM. Test leads usually include a fixed test probe at one end for connecting to the circuit and a banana plug at the other end for connecting to DMM banana jacks. Some test leads have connectors for detachable accessories such as test probes or test clips. This provides the convenience of using different accessories without having to change test leads. **See Figure 13-1.** Test leads should be very flexible and have double insulation that is resistant to cracking.

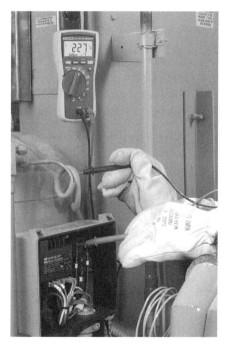

A magnetic hanging accessory offers convenience when taking and reading measurements.

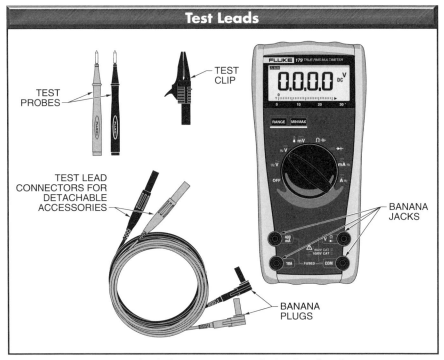

Figure 13-1. Some test leads have connectors which allow for changing accessories without changing test leads.

WARNING: Always inspect test leads for damaged insulation. Damaged test leads should be replaced immediately. Ensure that the test leads have good continuity without excessive resistance. Do not use damaged test leads to verify energized or de-energized circuits.

Test leads can be tested for continuity by setting the DMM to Resistance mode and touching the metal tips together. The measurement displayed is the resistance of the test leads. The resistance of test leads should range from 0.2 Ω to 0.5 Ω. If the DMM displays a higher resistance measurement, the test leads are damaged and should be replaced.

Test clips and test probes are used with test leads for a specific application. A *test clip* is a DMM accessory attached to a test lead used for making temporary connections for test measurements. **See Figure 13-2.** Test clips should include a safety grip that allows the test clip to be opened and closed at a safe distance from the connection point and/or an insulated cover. Noninsulated alligator clips should never be used for test connections.

Current flow in the grounding conductor is an indication that the neutral has been used for the grounding of equipment.

Test Clips	
Clip	**Description**
Industrial Test Clip 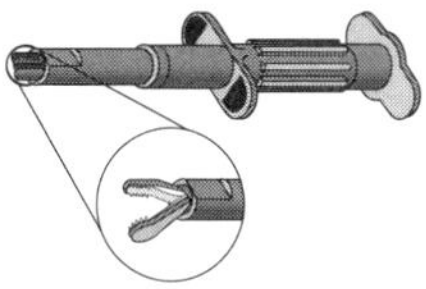	Insulated spring-loaded alligator clip for attaching to larger test points such as terminal screw or equipment ground
Hook-Style Test Clip 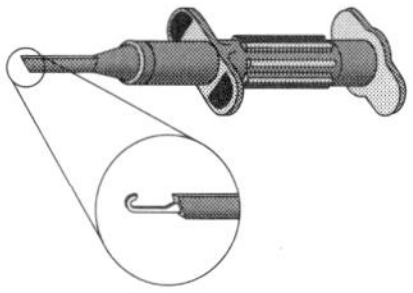	Insulated spring-loaded hook clip for attaching to smaller test points such as resistor lead wire and circuit components
Alligator Clip 	Insulated clip for attaching to test point or standard test probe to adapt for clamping function
Large Jaw Alligator Clip 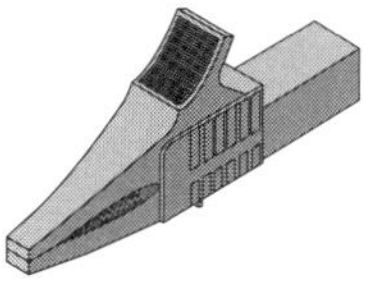	Insulated clip used for larger test point connections
Insulation Piercing Test Clip 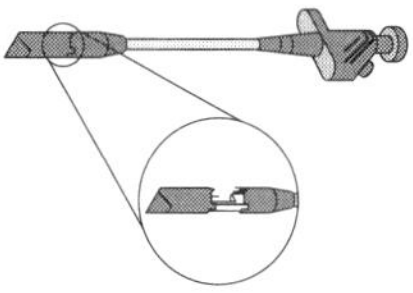	Insulated clip with spring-loaded probes for piercing insulated wire
Pin-Grabber Test Clip 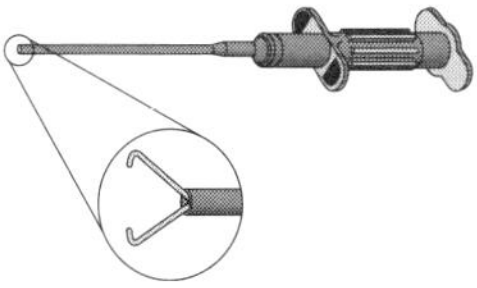	Insulated clip with spring-loaded pin-grabber hooks for test points in close contact areas

Figure 13-2. Test clips are designed for the specific connection required when taking test measurements.

A *test probe* is a DMM accessory used for making electrical contact for test measurements. Standard test probes have a sharp metal point attached to an insulated handle. The sharp metal point is used for making electrical contact by piercing insulation material or touching test points in tight places. **See Figure 13-3.** In industrial environments, both the test leads and test probes must be rated for both the electrical measurement category and voltage level for which they will be used.

Test probes should include insulated finger guards and safety grips to prevent accidental contact. Specialized test probes can increase the measurement range and/or capabilities to enhance DMM usefulness.

Clamp-On Current Probe Accessories

Clamp-on current probe accessories are used to take current readings without opening a circuit and to extend the current measurement range of the DMM. Current is measured in the circuit by measuring the strength of the magnetic field around a conductor. A clamp-on current probe accessory is preferred for taking most current measurements and is the safest method in high-energy electrical circuits or where shutting off the power to open the circuit is not practical. **See Figure 13-4.**

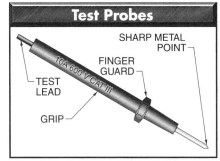

Test Probes

SHARP METAL POINT

FINGER GUARD

TEST LEAD

GRIP

10A 600 V CAT III

Figure 13-3. A test probe is used to make electrical contact by piercing insulation material or touching test points in tight places.

Clamp-On Current Probe Accessories	
Probe	**Description**
AC Clamp-On Current Probe Accessory ~AC	Used to measure AC currents only that are within the range of the clamp. Preferred method of measuring AC current, especially in high current measurements that cannot be interrupted.
AC/DC Clamp-On Current Probe Accessory ~ == AC DC	Used to measure AC or DC currents that are within the range of the clamp. Preferred method of measuring AC or DC current, especially in high current measurements and circuits that cannot be interrupted.

Figure 13-4. Clamp-on current probe accessories measure current in a circuit by measuring the strength of the magnetic field around a conductor.

High-Voltage Test Probes

A *high-voltage test probe* is a DMM accessory used to increase the voltage measurement range above the DMM listed range. A high-voltage test probe has a very high resistance which adds to the DMM input resistance circuit. This reduces the applied voltage to the DMM to approximately 1/1000 of the actual voltage present on the test probes measuring tip. For example, 25,000 V (25 kV) at the test probe tip can be reduced with a high-voltage test probe to approximately 25 V at the DMM jacks. **See Figure 13-5.**

High-voltage test probes are designed to take measurements in high-voltage, low-power applications such as TV picture tubes and electronic ignition systems. High-voltage test probes are not designed to take measurements in high-power applications such as high-voltage power distribution systems, induction-type heaters, X-ray equipment, and broadcast transmitters.

WARNING: High-voltage test probes are used with dangerously high voltages in low-power applications. To avoid electrical shock or damage, use high-voltage test probes only in dry conditions. Always use one hand when taking measurements, and wear approved protective rubber gloves and safety glasses.

 Silicon is the most widely used semiconductor material in solid-state components.

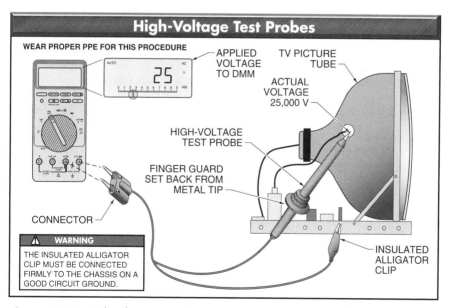

Figure 13-5. High-voltage test probes are used to increase the voltage measurement range of a DMM.

TEMPERATURE

Temperature is the measurement of the intensity of heat. *Heat* is thermal energy. Temperature is commonly measured in degrees Fahrenheit (°F) or degrees Celsius (°C). Converting one unit to the other is required because both Fahrenheit and Celsius are commonly used in the electrical field. **See Figure 13-6.**

To convert a Fahrenheit temperature reading to Celsius, subtract 32 from the Fahrenheit reading and divide by 1.8. To convert Fahrenheit to Celsius, apply the formula:

$$°C = \frac{\left(°F - 32\right)}{1.8}$$

where

°C = degrees Celsius

°F = degrees Fahrenheit

32 = difference between bases

1.8 = ratio between bases

To convert a Celsius temperature reading to Fahrenheit, multiply 1.8 by the Celsius reading and add 32. To convert Celsius to Fahrenheit, apply the formula:

$$°F = (1.8 \times °C) + 32$$

where

°C = degrees Celsius

°F = degrees Fahrenheit

32 = difference between bases

1.8 = ratio between bases

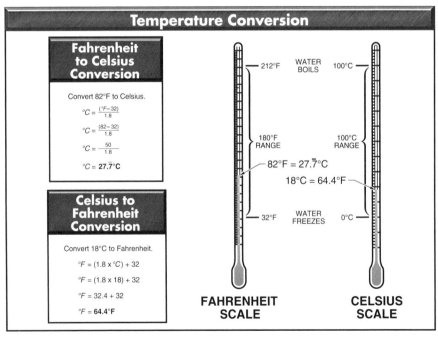

Figure 13-6. Temperature is commonly measured in degrees Fahrenheit (°F) or degrees Celsius (°C).

Temperature Probes

A *temperature probe* is a DMM accessory that measures the temperature of liquids, gases, surfaces, and pipes. The temperature probe required is based on the material to be measured, the temperature measurement range, and the accuracy required. DMM advanced features such as MIN MAX Recording mode and Relative mode can also be used for recording changing values.

Contact Temperature Probes. A *contact temperature probe* is a DMM accessory that measures temperature at a single point by direct contact with the area measured. Contact temperature probes can be used to measure the temperature of various solids, liquids, and gases depending on the probe used. Contact temperature probes can be connected directly to DMMs that now include on-board temperature measuring capabilities with temperature selector switch positions and direct temperature readouts. If the DMM does not have on-board temperature capabilities, a temperature module connected to the DMM can be used. When using a module, the DMM is typically set to measure mVDC. Most temperature modules deliver 1 mVDC for every degree in Fahrenheit or Celsius. **See Figure 13-7.**

WARNING: Never contact temperature probes for measurements on live circuits.

The most common contact temperature probe uses a thermocouple for temperature measurement. A *thermocouple* is a device that produces electricity when two different metals joined together are heated. The voltage produced is proportional to the measured temperature. The higher the temperature at the heated end, the higher the voltage produced at the other end. Voltage produced is typically limited to a few millivolts. A contact temperature probe with a thermocouple is connected to a temperature module inserted into the DMM. The voltage produced in mVDC is converted to a temperature measurement usually equivalent to 1 mV for every degree in Fahrenheit or Celsius.

WARNING: Use a temperature probe that is rated higher than the highest possible temperature to be measured, and wear proper protective equipment.

Noncontact Temperature Probes. A *noncontact temperature probe* is a DMM accessory used for taking temperature measurements on energized circuits or on moving parts. An *infrared temperature probe* is a noncontact temperature probe used for taking temperature measurements by sensing the infrared energy emitted by a material. All materials emit infrared energy in proportion to their temperature. Infrared temperature probes are used to take temperature measurements of electrical distribution systems, motors, bearings, and switching circuits. Although noncontact temperature probes can be added to a DMM to take temperature measurements, stand-alone IR (infrared) meters are becoming the preferred test instrument for taking temperature measurements.

Heat in an electrical device is usually caused by resistance resulting from a faulty electrical connection. The higher the resistance, the greater the amount of heat produced.

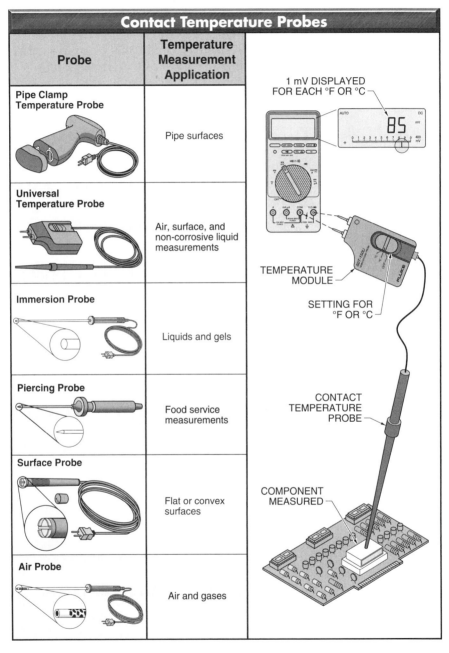

Contact Temperature Probes

Probe	Temperature Measurement Application
Pipe Clamp Temperature Probe	Pipe surfaces
Universal Temperature Probe	Air, surface, and non-corrosive liquid measurements
Immersion Probe	Liquids and gels
Piercing Probe	Food service measurements
Surface Probe	Flat or convex surfaces
Air Probe	Air and gases

1 mV DISPLAYED FOR EACH °F OR °C

TEMPERATURE MODULE

SETTING FOR °F OR °C

CONTACT TEMPERATURE PROBE

COMPONENT MEASURED

Figure 13-7. A contact temperature probe is a DMM accessory that measures temperature at a single point by direct contact with the area measured.

The temperature difference at a faulty electrical connection depends on the current flow and the resistance of the connection. Temperature measurement with a noncontact temperature probe and DMM with direct temperature measurement capability follows standard procedures. **See Figure 13-8.**

1. Wear proper PPE for area in which work is to be performed.

2. Connect noncontact temperature probe to the COM (common) and voltage/temperature jacks.

3. Set function switch to the temperature setting (mVDC setting). *Note:* Change meter function from mVDC to temperature, if required.

4. Take ambient temperature measurement for reference.

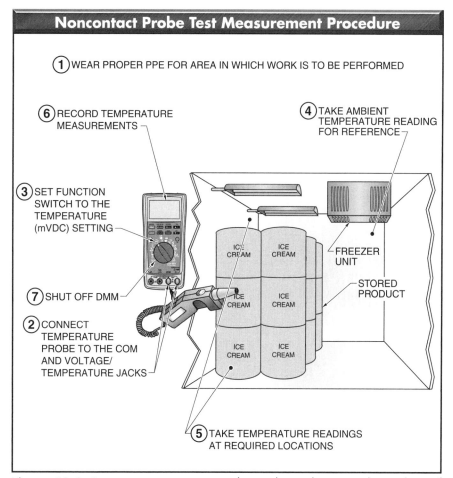

Noncontact Probe Test Measurement Procedure

①WEAR PROPER PPE FOR AREA IN WHICH WORK IS TO BE PERFORMED

⑥RECORD TEMPERATURE MEASUREMENTS

④TAKE AMBIENT TEMPERATURE READING FOR REFERENCE

③SET FUNCTION SWITCH TO THE TEMPERATURE (mVDC) SETTING

FREEZER UNIT

STORED PRODUCT

⑦SHUT OFF DMM

②CONNECT TEMPERATURE PROBE TO THE COM AND VOLTAGE/ TEMPERATURE JACKS

ICE CREAM

⑤TAKE TEMPERATURE READINGS AT REQUIRED LOCATIONS

Figure 13-8. A noncontact temperature probe can be used to assess the condition of electrical connections and operating equipment.

5. Take temperature measurements at required locations.

6. Record temperature measurements.

7. Shut off DMM.

Temperature measurement can also be used to assess equipment condition. In general, a temperature rise of 85°F above ambient temperature indicates a fault that requires routine maintenance. Routine maintenance is performed to remedy the problem before it causes permanent damage. A 100°F difference indicates a dangerous problem. Corrective action involves immediately shutting down the system and repairing the problem.

HIGH-FREQUENCY TEST PROBES

All DMMs can measure AC voltages, but the voltage must be within the listed frequency measurement range of the DMM. A *high-frequency test probe* is a DMM accessory that increases the frequency measurement range of the DMM. Frequency ranges vary depending on the application. **See Figure 13-9.**

High-Frequency Test Probe

WEAR PROPER PPE FOR THIS PROCEDURE

DMM FREQUENCY
RANGE 0.5 Hz – 200 kHz

Specifications

Function	Range	Resolution	Accuracy
Frequency	199.99	0.01 Hz	± (0.005% + 1)
(0.5 Hz – 200 kHz,	1999.9	0.1 Hz	± (0.005% + 1)
pulse width >2 μs)	19.999 kHz	0.001 kHz	± (0.005% + 1)
	199.99 kHz	0.01 kHz	± (0.005% + 1)
	>200 kHz	0.1 kHz	Unspecified

Frequency Ranges

Designation	Range
Video frequency	0 Hz – 4.5 MHz+
Power frequency	10 Hz – 1 kHz (50 Hz, 60 Hz common)
Human ear frequency	20 Hz – 20 kHz
Human voice frequency	300 Hz – 3 kHz
Very low frequency (VLF)	>3 kHz – 30 kHz
Low frequency (LF)	>30 kHz – 300 kHz
Medium frequency (MF)	>300 kHz – 3 MHz
High frequency (HF)	>3 MHz – 30 MHz
Very high frequency (VHF)	>30 MHz – 300 MHz

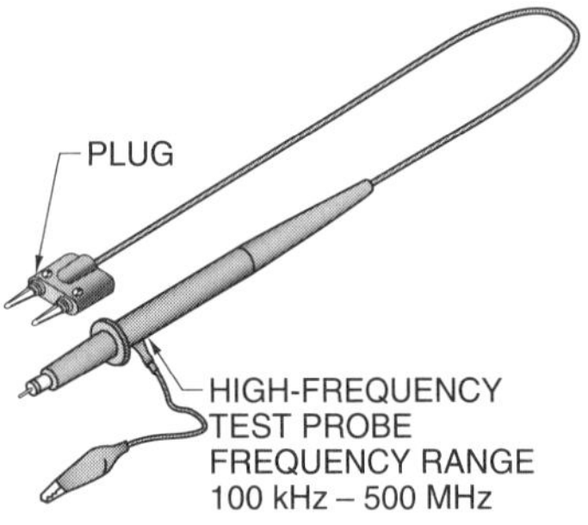

PLUG

HIGH-FREQUENCY
TEST PROBE
FREQUENCY RANGE
100 kHz – 500 MHz

Figure 13-9. A high-frequency test probe increases the frequency measurement range of the DMM.

Applications such as output signals from communication devices (transmitters) require a high-frequency test probe. Although the high-frequency test probes extend the DMM frequency measurement range, the voltage measurement is limited to the test probe rating.

PRESSURE/VACUUM MODULE

In many applications, measurement of system pressure or vacuum is required when troubleshooting or servicing the system. A *pressure/vacuum module* is a DMM accessory used for taking pressure or vacuum measurements. The pressure/vacuum module is plugged into the DMM, and a transducer is connected to the component or system. A *transducer* is a device used to convert physical parameters, such as temperature, pressure, and weight, into electrical signals. The pressure or vacuum present is sensed and sent to the pressure/vacuum module plugged into the DMM. See Figure 13-10.

Common applications of a pressure/vacuum module include taking hydraulic or pneumatic system pressure measurements up to 500 psi and vacuum measurements typically up to 29.99″ Hg. DMM advanced features such as MIN MAX Recording mode and Relative mode can also be used for monitoring changing values.

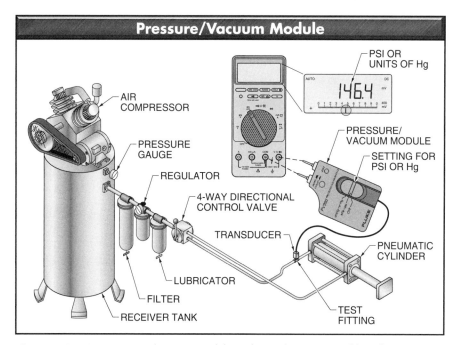

Figure 13-10. A pressure/vacuum module and transducer are used for taking pressure or vacuum measurements with a DMM.

FIBER-OPTIC METER

A *fiber-optic meter* is a DMM accessory used to test fiber-optic cable and fittings. *Fiber optics* is a technology that uses transmitted light signals through flexible fibers of glass, plastic, or other transparent materials. Fiber-optic cable can be used to connect two electronic circuits together. Electronic circuits convert electrical signals into light signals at the fiber-optic transmitter or convert the light signals back into electronic signals at the fiber-optic receiver.

When transmitting light through a fiber-optic cable, light loss must be kept to a minimum. A fiber-optic meter is used with some DMMs to test fiber-optic cable and fittings for any losses. **See Figure 13-11.** For example, light loss at a coupler can be detected by first taking a measurement at the cable before the coupler. With the DMM set for Relative mode, a measurement is taken at the coupler. Light loss is indicated by the difference in measurement displayed.

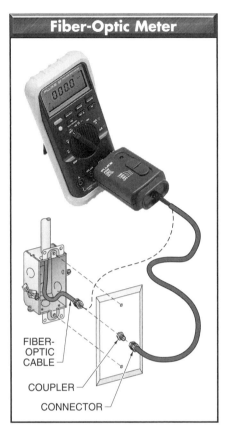

Figure 13-11. A fiber-optic meter is a DMM accessory used to measure light loss in fiber-optic cable and fittings.

Test probes must be capable of making contact only with the component tested.

CARBON MONOXIDE PROBE

A *carbon monoxide (CO) probe* is a DMM accessory used to measure the amount (level) of CO in the air. *Carbon monoxide (CO)* is a colorless and odorless toxic gas that is produced when substances that contain carbon (wood, natural gas, gasoline, fuel oil, etc.) are incompletely burned. A self-contained CO meter can also be used to measure the amount (level) of CO in the air.

High levels of carbon monoxide (CO) can be deadly. A CO probe accessory or CO meter measures the level of CO in parts per million (PPM). Carbon monoxide level standards are set by governing agencies. **See Figure 13-12.**

A carbon monoxide probe can be used to detect leaks from malfunction or improper installation.

Carbon Monoxide (CO) Level Standards*	
0–1 PPM	Normal background levels
9 PPM	ASHRAE Standard 62-1989 for living areas
50 PPM	OSHA enclosed space 8-hour average level[†]
100 PPM	OSHA exposure limit[†]
200 PPM	Mild headache, fatigue, nausea, and dizziness
800 PPM	Dizziness, nausea, and convulsions. Death within 2 to 3 hours

[†] U.S. Department of Labor, Occupational Safety and Health Administration (OSHA) Regulation 1917.24: The CO content in any enclosed space shall be maintained at not more than 50 PPM (0.005%). Remove employees from enclosed space if the CO concentration exceeds 100 PPM (0.01%).

* in parts per million (ppm)

Figure 13-12. Carbon monoxide levels are measured in parts per million (PPM).

Common sources of CO emissions include the following:

- Malfunctioning or improperly installed gas/oil/kerosene furnaces and heaters or fireplaces.

- Obstructed or undersized chimneys or flue exhausts.

- Poorly maintained gas, oil, or kerosene appliances.

- Improperly ventilated combustion engines (automobiles, portable generators, lawnmowers, etc.)

For maximum safety, CO measurements should be taken when working on or around any CO producing device. **See Figure 13-13.**

Carbon Monoxide (CO) Emission Sources		
Appliance	**Fuel**	**Typical Problems**
Gas furnaces Room heaters	Oil, natural gas, or LPG (liquefied petroleum gas)	1. Cracked heat exchanger 2. Not enough air to burn fuel properly 3. Defective/blocked flue 4. Maladjusted burner 5. Building not properly pressurized
Central heating furnaces	Coal or kerosene	1. Cracked heat exchanger 2. Not enough air to burn fuel properly 3. Defective grate
Room heaters Central heaters	Kerosene	1. Improper adjustment 2. Wrong fuel 3. Wrong wick or wick height 4. Not enough air to burn fuel 5. System not properly vented
Water heaters	Natural gas or LPG	1. Not enough air to burn fuel properly 2. Maladjusted burner 3. Misuse as a room heater 4. System not properly vented
Ranges Ovens	Natural gas or LPG	1. Not enough air to burn fuel properly 2. Maladjusted burner 3. Misuse as a room heater 4. System not properly vented
Stoves Fireplaces	Gas, wood, coal	1. Not enough air to burn fuel properly 2. Defective/blocked flue 3. Green or treated wood 4. Cracked heat exchanger 5. Cracked firebox

Figure 13-13. Carbon monoxide is a colorless and odorless toxic gas produced when substances that contain carbon are incompletely burned.

DMM FEATURES

DMMs vary in cost based on operational features, safety features, and specifications. DMM required features are determined by the anticipated applications in the field. DMMs with the minimum required features for the application are commonly selected for economy. However, DMMs with more than the minimum required features often provide a good investment for the technician. Additional DMM features provide greater capabilities and usefulness if testing and troubleshooting tasks increase in complexity. In addition, other applications and/or changes in testing standards may require greater precision. DMM manufacturers provide information about specific DMM features and usage.

Resolution

Resolution is the degree of precise measurement a DMM is capable of making. **See Figure 14-1.** For example, if the DMM has a resolution of 1 mV on the 3 V range, it is possible to see a change of 1 mV while reading 1 V. Resolution may be listed in the DMM specifications as maximum resolution. *Maximum resolution* is the smallest value that can be discerned on the lowest range setting.

For example, a maximum resolution of 100 mV (0.1 V) means that when the range is set to measure the highest possible voltage, the displayed voltage is rounded to the nearest tenth of a volt.

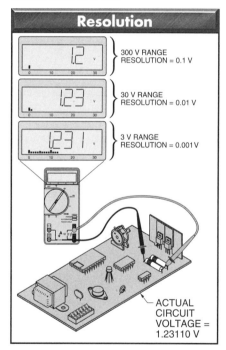

Figure 14-1. Resolution is the degree of precise measurement a DMM is capable of making.

Resolution is improved by reducing the range setting of the DMM as long as the measurement is within the set range. Most autoranging DMMs automatically select the range that provides the best resolution for the measurement. If the measurement is higher than the set range, the DMM displays OL (overload). The most accurate measurement is obtained at the lowest possible range setting without overloading the DMM.

To improve resolution for better accuracy when measuring low DC voltage, some DMMs include a lower DC voltage function switch setting. For example, a DMM may include a 600 mV DC setting in addition to the VDC setting. Caution is required when using the 600 mV setting because voltage over 600 mV (0.6 V) produces an overload. DMM range and resolution are related and are sometimes specified together in DMM specifications. **See Figure 14-2.**

Range and Resolution

Range	Resolution
300.0 mV	0.1 mV (0.0001 V)
3.000 V	1 mV (0.001 V)
30.00 V	10 mV (0.01 V)
300.0 V	100 mV (0.1 V)
1000 V	1000 mV (1 V)

Figure 14-2. DMM range and resolution are related and are sometimes listed together in DMM specifications.

Digits and Counts

Digits and counts are terms used to describe DMM resolution. DMMs are grouped by the number of digits or counts displayed. For example, a standard 3½ digit DMM can display three full digits and a half digit. The three full digits display a number from 0 to 9. The half digit displays a 1 or is left blank. The three full digits are the least significant digits, and the half digit is the most significant digit. The largest number on a standard 3½ digit display is 1999. **See Figure 14-3.**

DMM resolution is also specified in counts. For example, a 3200 count DMM can display any number from 0 to 3199, and a 4000 count DMM can display any number from 0 to 3999. DMMs with more counts offer better resolution for certain measurements. For example, a 1999 count DMM cannot measure down to a tenth of a volt if measuring 200 V or more. The largest number on a 3½ digit display is 1999. However, many modern 3½ digit DMM models can display up to 5999.

Accuracy

Accuracy is the largest allowable error that occurs under specific operating conditions. Accuracy is expressed in percent and indicates how close the displayed measurement is to the actual value of the signal measured. The accuracy of a specific DMM is less or more important depending on the application. For example, most AC power line voltages vary ±5% or more. An example of this variation is a voltage measurement taken at a standard 115 VAC receptacle. If a DMM is only used to check if a receptacle is energized, a DMM with a ±3% measurement accuracy is appropriate.

Some applications, such as calibration of automotive, medical, aviation, or specialized industrial equipment, may require more accuracy. A reading of 100.0 V on a DMM with an accuracy of ±2% can range from 98.0 to 102.0 V. Accuracy may also include a specified amount of digits (counts) added to the basic accuracy rating. For example, an accuracy of ±(2%+2) means that a reading of 100.0 V on the DMM can be from 97.8 V to 102.2 V. Use of a DMM with higher accuracy allows a greater number of applications.

True rms and Average-Responding rms

DMMs are specified as true rms (root-mean-square) or average-responding rms indicating. Both true rms and average-responding rms indicating DMMs can accurately measure standard sinusoidal (pure AC) waveforms.

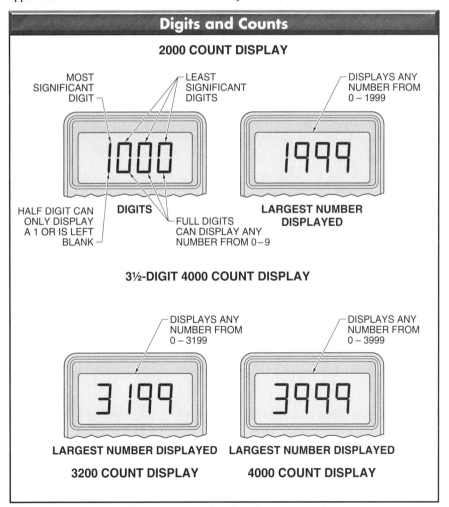

Figure 14-3. Digits and counts are used to describe DMM resolution.

Only a true-rms DMM can obtain accurate measurements on nonsinusoidal waveforms such as square waveforms, sawtooth waveforms, and distorted sinusoidal waveforms. In applications where AC measurements are taken on circuits that contain sinusoidal waveforms, an average-responding DMM works well.

As more circuits include variable-speed motor drives, electronic ballasts, and computers, the possibility of nonsinusoidal waveforms increases. In addition, a true-rms DMM is the best choice for taking measurements on power lines where AC waveform characteristics are unknown. A DMM that does not specify that it is true rms is usually average responding. Because more voltages and currents are comprised of a distorted wave shape and/or include both AC and DC, some DMM models have an AC/DC function that can capture both the AC and DC values of a complex waveform and display the true rms value.

Response Time

When using the MIN MAX Recording mode for circuit values, the rated response time affects measurement accuracy. *Response time* is the length of time an input must stay at a new value to be recorded. DMMs can have a set response time or a changeable response time. The best response time depends upon the application. The fastest response time is not always the best. For example, if a transient voltage is to be captured, a fast response time such as 1 ms is best. **See Figure 14-4.**

A slower response time such as 1 sec is best when recording temperature change. DMMs commonly have response times listed as 1 ms, 100 ms, and/or 1 sec. Functions with response times of 1 ms or faster are commonly referred to as "Peak MIN MAX" functions, as usually indicated on the meter's function switch.

DMM APPLICATIONS

Electrical systems vary in complexity, currents, and voltages. Simple problems such as blown fuses occur in any system and can be isolated using a DMM with basic features.

Complex problems in an electrical system require a DMM with advanced features and/or accessories for measuring temperature, pressure/vacuum, and higher voltages, currents, and frequencies. DMM applications can be grouped into automotive, HVAC, electrical maintenance, or electronic circuit applications. In each group, basic testing procedures and/or system troubleshooting procedures commonly require certain DMM features. **See Figure 14-5.**

Automotive Applications

Electrical systems on automobiles have evolved from basic electrical systems such as the starting, ignition, and charging systems. With advances in technology, other automotive electrical systems have been developed for passenger comfort, safety, and emission controls. For example, passenger comfort systems require circuits for heating and air conditioning, power seats, and power mirrors. Safety systems require circuits for air bag operation, anti-lock brakes, and traction control. Emission-control systems require circuits for electronic fuel injection, ignition control, and exhaust gas analysis. A computer system is commonly used to control many of these electrical systems.

Basic automotive troubleshooting tasks involve testing for blown fuses, taking voltage readings, and/or checking the resistance

of switches. A DMM capable of continuity testing, measuring DC voltage, and measuring resistance is adequate for these tasks. For more advanced automotive trouble-shooting tasks, a DMM with MIN MAX Recording mode, bar graph display, Diode Test mode, Capacitance Measurement mode, and Duty Cycle mode is used. The DMM should also accept accessories for measuring temperature, pressure/vacuum, and higher AC/DC currents.

Response Times

MIN MAX Recording Mode Application	Time		
	≤1 ms	100 ms	1 sec
Detect intermittent loose connections		X	
Capture voltage sags when starting high-power loads		X	
Capture starting voltage of motors started using reduced-voltage starting		X	
Record inrush currents		X	
Monitor output voltage from battery charger			X
Record temperature changes			X
Capture transient voltages on power lines	X		
Determine if circuit is overloaded at any time			X
Measure peak currents lasting 1 ms or longer	X		
Monitor frequency changes on power lines			X
Monitor if load is turned OFF			X

Figure 14-4. The preferred DMM response time is determined by the application.

DMM Applications and Features

Recommended Features	Automotive Basic Testing	Automotive System Troubleshooting	HVAC Basic Testing	HVAC System Troubleshooting	Electrical Maintenance Basic Testing	Electrical Maintenance System Troubleshooting	Electronic Circuit Basic Testing	Electronic Circuit System Troubleshooting
Bar Graph Display		X		X	X	X	X	X
Capacitance Measurement Mode		X		X		X	X	X
Continuity Test Mode	X	X	X	X	X	X	X	X
Diode Test Mode	X	X				X	X	X
Duty Cycle Mode		X						X
Frequency Counter Mode		X				X	X	X
MIN MAX Recording Mode		X		X		X		X
Peak MIN MAX		X		X	X	X	X	X
Relative Mode	X	X	X	X	X	X	X	X
Touch Hold® Mode	X	X	X	X	X	X	X	X
Temperature Measurement		X	X	X		X		X
LoZ			X	X	X	X	X	
LoOhms	X	X		X		X	X	X
LoVAC			X	X	X	X		

Figure 14-5. Different applications commonly require certain DMM features.

HVAC Applications

HVAC systems provide comfort to occupants in residential, commercial, and industrial buildings. HVAC systems range from basic resistant electric heaters to large computer-controlled systems. For example, adjustable frequency drives for air flow control, programmable thermostats, and/or electronic filters can be linked and computer controlled in a total energy management system.

For testing and troubleshooting basic HVAC problems, a DMM capable of measuring AC voltage, DC voltage, resistance, and continuity testing is adequate. In addition, a temperature measurement module and an AC clamp-on current probe accessory are also commonly used for basic troubleshooting tasks.

For more advanced HVAC troubleshooting tasks, a DMM with MIN MAX Recording mode, Relative mode, bar graph display, Diode Test mode, and Capacitance Measurement mode should be used. The DMM should also accept accessories for measuring temperature, pressure/vacuum, or higher AC current.

For common tasks such as testing voltage levels, testing fuses, measuring line current, and testing switches and conductors for continuity, a basic DMM is adequate. For more advanced electrical maintenance troubleshooting tasks on adjustable frequency drives, programmable controllers, and solid-state controls, a DMM with MIN MAX Recording mode, Relative mode, bar graph display, Diode Test mode, Frequency Counter mode, and Capacitance Measurement mode should be used. The DMM should also accept accessories for measuring temperature, pressure/vacuum, and higher AC and DC current.

Electronic Circuit Applications

Electronic circuits as with other circuits can develop simple problems such as a blown fuse. However, electronic circuits commonly involve more complex problems. A DMM having MIN MAX Recording mode, Relative mode, bar graph display, Diode Test mode, Frequency Counter mode, and Capacitance Measurement mode should be used. Electronic circuit measurements usually require a DMM with a high degree of measurement accuracy. Electronic circuits may also require the use of high-voltage measurement or fiber optic testing accessories. For example, high-voltage measurements may be required when testing television equipment. Fiber optic measurements may be required when testing communication equipment.

DMM temperature measurement accessories can be used for troubleshooting tasks in automotive applications.

Electrical Maintenance Applications

Electrical maintenance includes testing and troubleshooting power supply and load control in industrial and residential circuits.

DMM SPECIFICATIONS

A *DMM specification* is information detailing the capabilities and limitations of a specific DMM. Specifications are provided by the DMM manufacturer and include general and measuring specifications. **See Figure 14-6.** General specifications provide information such as display response time, operating temperature, and storage temperature. For example, display response time listed for a specific DMM indicates that the bar graph display updates the measurement 10 times faster than the digital display (25 times per second compared to 2.5 times per second). General specifications also state response times for voltage and resistance measurements.

Measuring specifications list the measuring accuracy of the DMM. For example, DMM measuring specifications for AC or DC voltage, resistance, and AC or DC current are listed. When the DMM is set to 6000.0 Ω resistance range, the measurement accuracy is ±(0.2% +1). This accuracy means that if a DMM is measuring a 1000 Ω resistance, the DMM reading could range from 998.8 Ω to 1001.2 Ω.

In addition to listing the measuring accuracy, specifications also list the measurement range. For example, a DMM can measure both AC and DC voltages up to 1000 V and resistances up to 500 MΩ.

Trend Capturing (Logging)

Recording measurements over time is a valuable feature when troubleshooting

or evaluating how a circuit or piece of electrical equipment is operating over a specified time period. Digital multimeters with a trend capturing feature record measurements (voltage, current, resistance, temperature, etc.) over units of time (seconds, minutes, or hours) and plot the measurements as a single line. **See Figure 14-7.**

Being able to view electrical measurements over time is helpful when troubleshooting or checking a circuit load. For example, being able to plot the amount of current used over a given time period of 24 hours can provide the peak, average, and low current readings. When connected to a 15 A branch circuit (or 200 A main line), a plot of current over time indicates if additional loads can be safely added. If the measurements indicate that the 15 A branch circuit never draws more than 8 A, then additional loads can be added.

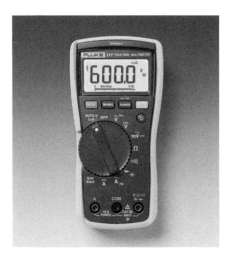

Measuring specifications list the measuring accuracy of AC or DC voltage, resistance, and AC or DC current of a DMM.

DMM Specifications

General Specifications

True-rms for voltage and current
4½ digit mode for precise measurements (20,000 counts)
Measure up to 1000 V AC and DC
Built-in thermometer
Resistance, continuity, and diode test
10,000 µF capacitance range

Measuring Specifications

Voltage DC	Maximum Voltage: Accuracy: Maximum Resolution:	1000 V Fluke 83 V: ±(0.1% +1) Fluke 87 V: ±(0.05% +1) +1 Fluke 83 V: 100 µV Fluke 87 V: 10 µV
Voltage AC	Maximum Voltage: Accuracy: AC Bandwidth Maximum Resolution:	1000 V Fluke 83 V: ±(0.5% +2) Fluke 87 V: ±(0.7% +2) true-rms Fluke 83 V: 5kHz Fluke 87 V: 20kHz* *with low pass filter; 3 db @ 1 kHz 0.1 mV
Current AC	Maximum Amps: Amps Accuracy: Maximum Resolution:	10 A (20 A for 30 seconds maximum) Fluke 83 V: ±(1.2% +2) Fluke 87 V: ±(1.0% +2) true-rms 0.1 µA
Resistance	Maximum Resistance: Accuracy: Maximum Resolution:	50 MΩ Fluke 83 V: ±(0.4% +1) Fluke 87 V: ±(0.2% +1) +1 0.1 Ω

Figure 14-6. The capabilities and limitations for specific DMMs are given in DMM specifications.

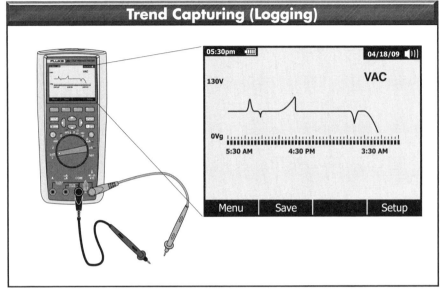

Figure 14-7. Digital multimeters with a trend capturing feature record measurements over units of time (seconds, minutes, or hours) and plot the measurements as a single line.

Selected DMM Abbreviations

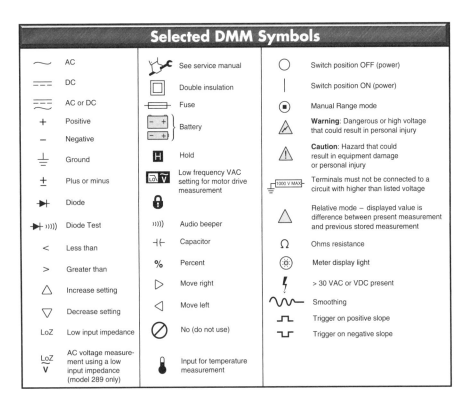

AC	Alternating current or voltage
DC	Direct current or voltage
V	Volts
mV	Millivolts
kV	Kilovolts
A	Amperes
mA	Milliamperes
μA	Microamperes
W	Watts
kΩ	Kilohms
MΩ	Megohms
Hz	Hertz
kHz	Kilohertz
μF	Microfarads
nF	Nanofarads
°F	Degrees Fahrenheit
°C	Degrees Celsius
LEAD	Indicates test lead position does not match function switch position
LOG	Readings are being recorded
Lo	Low
nS	nanosiemens (1×10-9 or 0.000000001 Siemens)
MEM	Memory
MS	Time display in minutes:seconds
HM	Time display in hours:minutes

RPM	Revolutions per minute
COM	Common
OL	Overload
T	Time
LSD	Least significant digit
MAX	Maximum
MIN	Minimum
AVG	Average
TRIG	Trigger
V_{avg}	Average voltage
V_{max}	Peak voltage
V_{p-p}	Peak-to-peak voltage
V_{rms}	Root-mean-square (rms) voltage
Hi-Z	High input impedance
dB	Decibel
dBV	Decibel volts
dBW	Decibel watts

Selected DMM Symbols

∿	AC	⚡	See service manual	◯	Switch position OFF (power)
===	DC	▢	Double insulation	|	Switch position ON (power)
≝	AC or DC	⊏▭	Fuse	◉	Manual Range mode
+	Positive	🔋	Battery	⚠	**Warning**: Dangerous or high voltage that could result in personal injury
−	Negative			⚠	**Caution**: Hazard that could result in equipment damage or personal injury
⏚	Ground	**H**	Hold		
±	Plus or minus	🔲	Low frequency VAC setting for motor drive measurement	⚡1000 V MAX	Terminals must not be connected to a circuit with higher than listed voltage
▸⊢	Diode	🔒			
▸⊢)))	Diode Test	)))	Audio beeper	△	Relative mode – displayed value is difference between present measurement and previous stored measurement
<	Less than	⊣⊢	Capacitor	Ω	Ohms resistance
>	Greater than	%	Percent	⊙	Meter display light
△	Increase setting	▷	Move right	⚡	> 30 VAC or VDC present
▽	Decrease setting	◁	Move left	∿∿	Smoothing
LoZ	Low input impedance	⊘	No (do not use)	⊓	Trigger on positive slope
LoZ/V	AC voltage measurement using a low input impedance (model 289 only)	🌡	Input for temperature measurement	⊔	Trigger on negative slope

DMM Terminology . . .

Term	Symbol	Definition
AC		Continually changing current that reverses direction at regular intervals; standard U.S. frequency is 60 Hz
AC COUPLING		Signal that passes an AC signal and blocks a DC signal; used to measure AC signals that are riding on a DC signal
ACCURACY ANALOG METER		Largest allowable error (in percent of full scale) made under normal operating conditions; the reading of a DMM set on the 250 V range with an accuracy rating of ± 2% could vary ± 5 V; analog DMMs have greater accuracy when readings are taken on the upper half of the scale
ACCURACY DIGITAL METER		Largest allowable error (in percent of reading) made under normal operating conditions; a reading of 100.0 V on a DMM with an accuracy of ±2% is between 98.0 V and 102.0 V; accuracy may also include a specified amount of digits (counts) that are added to the basic accuracy rating; for example, an accuracy of ±2% (±2 digits) means that a display reading of 100.0 V on the DMM is between 97.8 V and 102.2 V
AC/DC		Indicates ability to read or operate on alternating and direct current
AC FREQUENCY RESPONSE		Frequency range over which AC voltage measurements are accurate
ALLIGATOR CLIP		Long-jawed, spring-loaded, insulated clamp used to safely make temporary electrical connections
AMBIENT TEMPERATURE		Temperature of air surrounding a DMM or equipment to which the DMM is connected
AMMETER		DMM that measures electric current
AMMETER SHUNT		Low-resistance conductor that is connected in parallel with the terminals of an ammeter to extend the range of current values measured by the ammeter
AMPLITUDE		Measure of AC signal alternation expressed in values such as peak or peak-to-peak
ATTENUATION		Decrease in amplitude of a signal
AUDIBLE		Sound that can be heard
AUTOHOLD		Function that captures a measurement, beeps, and locks the measurement on the digital display for later viewing and will automatically update with new reading
AUTORANGE MODE		Function that automatically selects a DMM's range based on signals received
AVERAGE VALUE		Value equal to 0.637 times the amplitude of a measured value

. . . DMM Terminology . . .

Term	Symbol	Definition
BACKLIGHT		Light that brightens the DMM display
BANANA JACK		DMM jack that accepts a banana plug
BANANA PLUG		Long, thick terminal connection on one end of a test lead used to make a connection to a DMM
BATTERY SAVE		Feature that enables a DMM to shut down when battery level is too low or no key is pressed within a set time
BNC		Coaxial-type input connector used on some DMMs
CAPTURE		Function that records and displays measured values
CELSIUS	$°C$	Temperature measured on a scale for which the freezing point of water is 0° and the boiling point is 100°
CLOSED CIRCUIT		Circuit in which two or more points allow a predesigned current to flow
CONTINUITY CAPTURE		Function used to detect intermittent open and short circuits as brief as 250 μs
COUNTS		Unit of measure of DMM resolution; a 1999 count DMM cannot display a measurement of $1/10$ of a volt when measuring 200 V or more; a 3200 count DMM can display a measurement of $1/10$ of a volt up to 320 V
CREST FACTOR		Ratio of peak value to the rms value; the higher the DMM crest factor, the wider the range of waveforms it can measure; in a pure sine wave, the crest factor is 1.41
db READOUT		Decibels (dB) unit of measure used to express the ratio between two quantities such as the gain or loss of amplifiers, filters, or attenuators in telecommunications or audio applications
DC		Current that constantly flows in one direction
DECIBEL (dB)		Measurement that indicates voltage or power comparison in a logarithmic scale
DIGITS		Indication of the resolution of a DMM; standard DMM with a 3½ digit specification can display three full digits on the right of the display (0 to 9) and ½ digit (1 or left blank) on the left (up to 1999 counts); newer DMMs, such as 6000, 20,000, or 50,000 count DMMs, use counts to more accurately specify resolution
DIODE		Semiconductor that allows current to flow in only one direction
DISCHARGE		Removal of an electric charge
DUAL DISPLAY		Feature that allows two separate waveforms to be displayed simultaneously
EARTH GROUND		Reference point that is directly connected to ground

. . . DMM Terminology . . .

Term	Symbol	Definition
EFFECTIVE VALUE		Value equal to 0.707 of the peak in a sine wave
FAHRENHEIT	°F	Temperature measured on a scale for which the freezing point of water is 32° and the boiling point is 212°
FREQUENCY		Number of complete cycles occurring per unit of time
FUNCTION SWITCH		Switch that selects the function (AC voltage, DC voltage, etc.) that a DMM is to measure
GLITCH		Momentary spike in a waveform
GLITCH DETECT		Function that increases the DMM sampling rate to maximize the detection of the glitch(es)
GROUND		Common connection to a point in a circuit whose potential is taken as zero
SAVE		Function that allows a measurement to be saved and stored
HARMONICS		Currents generated by electronic devices (nonlinear loads), which draw current in short pulses, not as a smooth sine wave; harmonic currents are whole-number multiples of the fundamental current (typically 60 Hz)
HOLD BUTTON	HOLD Ⓗ	Button that allows a DMM to capture and hold a stable measurement
INPUT ALERT		Function that provides an audible warning if test leads are in current input jacks but the function switch is not in amps position
LIQUID CRYSTAL DISPLAY (LCD)		Display that uses liquid crystals to display waveforms, measurements, and text on its screen
LoZ		Setting for low-impedance voltages; designed to prevent false ghost voltage readings/displays
MEASURING RANGE		Minimum and maximum quantity that a DMM can safely and accurately measure
MIN MAX		Function that captures and stores the highest and lowest measurements for later viewing; function can be used with any DMM measurement function such as volts, amps, etc.
MIN MAX INSTANTANEOUS PEAK		High-speed response time used to capture MIN MAX readings of a waveform peak value; can be used for crest factor calculations or to capture transient voltage or momentary voltage surge measurements
NOISE		Unwanted extraneous electrical signals
OPEN CIRCUIT		Circuit in which two (or more) points do not provide a path for current flow
OVERLOAD	OL	Condition of a DMM that occurs when a quantity to be measured is greater than the quantity the DMM can safely handle for the DMM range setting or display
PEAK		Maximum value of positive or negative alternation in a sine wave

. . . DMM Terminology

Term	Symbol	Definition
PEAK-TO-PEAK		Value measured from the maximum negative to the maximum positive alternation in a sine wave
POLARITY		Orientation of the positive (+) and negative (–) side of direct current or voltage
PROBE		Pointed metal tip of a test lead used to make contact with the circuit under test
PULSE		Waveform that increases from a constant value, then decreases to its original value
PULSE TRAIN		Repetitive series of pulses
RANGE		Quantities between two points or levels
RECALL		Function that allows stored information (or measurements) to be displayed
RECORD		Allows measurements to be recorded
RESOLUTION		Degree of measurement precision of DMM when taking measurement
RISING SLOPE		Part of a waveform displaying a rise in voltage
ROOT-MEAN-SQUARE		Value equal to 0.707 of the amplitude of a measured value
SAMPLE		Momentary reading taken from an input signal
SAMPLING RATE		Number of readings taken on a signal over time
SHORT CIRCUIT		Two or more points in a circuit that allow an unplanned current flow
TERMINAL		Point to which DMM test leads are connected
TERMINAL VOLTAGE		Voltage level that DMM terminals can safely handle
TREND CAPTURE		Function that allows a measurement to be recorded over time and displayed in a straight line
TRIGGER LEVEL		Fixed level at which DMM counter is triggered
WAVEFORM		Pattern defined by an electrical signal
ZOOM		Function that allows a waveform (or part of waveform) to be magnified

Common Prefixes

Symbol	Prefix	Equivalent
G	giga	1,000,000,000
M	mega	1,000,000
k	kilo	1000
base unit	—	1
m	milli	0.001
μ	micro	0.000001
n	nano	0.000000001
p	pico	0.000000000001
Z	impedance	ohms – Ω

Metric Prefixes

Multiples and Submultiples	Prefixes	Symbols	Meaning
$1,000,000,000,000 = 10^{12}$	tera	T	trillion
$1,000,000,000 = 10^{9}$	giga	G	billion
$1,000,000 = 10^{6}$	mega	M	million
$1000 = 10^{3}$	kilo	k	thousand
$100 = 10^{2}$	hecto	h	hundred
$10 = 10^{1}$	deka	d	ten
Unit $1 = 10^{0}$			
$.1 = 10^{-1}$	deci	d	tenth
$.01 = 10^{-2}$	centi	c	hundredth
$.001 = 10^{-3}$	milli	m	thousandth
$.000001 = 10^{-6}$	micro	μ	millionth
$.000000001 = 10^{-9}$	nano	n	billionth
$.000000000001 = 10^{-12}$	pico	p	trillionth

Initial Units	Final Units										
	giga	mega	kilo	hecto	deka	base unit	deci	centi	milli	micro	nano	pico
giga		3R	6R	7R	8R	9R	10R	11R	12R	15R	18R	21R
mega	3L		3R	4R	5R	6R	7R	8R	9R	12R	15R	18R
kilo	6L	3L		1R	2R	3R	4R	5R	6R	9R	12R	15R
hecto	7L	4L	1L		1R	2R	3R	4R	5R	8R	11R	14R
deka	8L	5L	2L	1L		1R	2R	3R	4R	7R	10R	13R
base unit	9L	6L	3L	2L	1L		1R	2R	3R	6R	9R	12R
deci	10L	7L	4L	3L	2L	1L		1R	2R	5R	8R	11R
centi	11L	8L	5L	4L	3L	2L	1L		1R	4R	7R	10R
milli	12L	9L	6L	5L	4L	3L	2L	1L		3R	6R	9R
micro	15L	12L	9L	8L	7L	6L	5L	4L	3L		3R	6R
nano	18L	15L	12L	11L	10L	9L	8L	7L	6L	3L		3R
pico	21L	18L	15L	14L	13L	12L	11L	10L	9L	6L	3L	

Metric Conversions

R = move decimal point one place to the right
L = move decimal point one place to the left

Ohm's Law

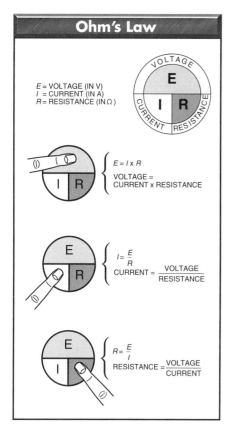

E = VOLTAGE (IN V)
I = CURRENT (IN A)
R = RESISTANCE (IN Ω)

$E = I \times R$

VOLTAGE =
CURRENT x RESISTANCE

$I = \dfrac{E}{R}$

CURRENT = $\dfrac{\text{VOLTAGE}}{\text{RESISTANCE}}$

$R = \dfrac{E}{I}$

RESISTANCE = $\dfrac{\text{VOLTAGE}}{\text{CURRENT}}$

Power Formula

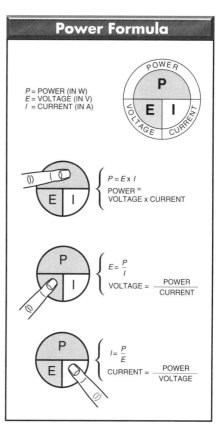

P = POWER (IN W)
E = VOLTAGE (IN V)
I = CURRENT (IN A)

$P = E \times I$

POWER =
VOLTAGE x CURRENT

$E = \dfrac{P}{I}$

VOLTAGE = $\dfrac{\text{POWER}}{\text{CURRENT}}$

$I = \dfrac{P}{E}$

CURRENT = $\dfrac{\text{POWER}}{\text{VOLTAGE}}$

Ohm's Law and Power Formula

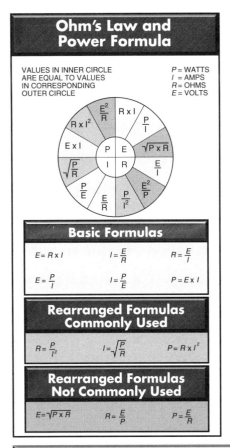

VALUES IN INNER CIRCLE ARE EQUAL TO VALUES IN CORRESPONDING OUTER CIRCLE

P = WATTS
I = AMPS
R = OHMS
E = VOLTS

Basic Formulas

$E = R \times I$ $I = \dfrac{E}{R}$ $R = \dfrac{E}{I}$

$E = \dfrac{P}{I}$ $I = \dfrac{P}{E}$ $P = E \times I$

Rearranged Formulas Commonly Used

$R = \dfrac{P}{I^2}$ $I = \sqrt{\dfrac{P}{R}}$ $P = R \times I^2$

Rearranged Formulas Not Commonly Used

$E = \sqrt{P \times R}$ $R = \dfrac{E}{P}$ $P = \dfrac{E}{R}$

Voltage, Current, and Impedance Relationship

E = VOLTAGE (IN V)
I = CURRENT (IN A)
Z = IMPEDANCE (IN Ω)

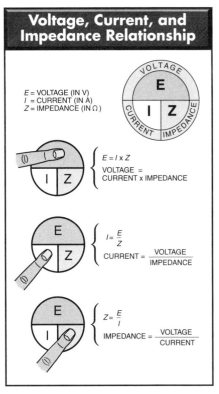

$E = I \times Z$
VOLTAGE = CURRENT x IMPEDANCE

$I = \dfrac{E}{Z}$
CURRENT = $\dfrac{\text{VOLTAGE}}{\text{IMPEDANCE}}$

$Z = \dfrac{E}{I}$
IMPEDANCE = $\dfrac{\text{VOLTAGE}}{\text{CURRENT}}$

Horsepower Formulas

To Find	Use Formula	Example		
		Given	Find	Solution
HP	$HP = \dfrac{I \times E \times E_{ff}}{746}$	240 V, 20 A, 85% E_{ff}	HP	$HP = \dfrac{I \times E \times E_{ff}}{746}$
				$HP = \dfrac{240 \text{ V} \times 20 \text{ A} \times 85\%}{746}$
				$HP = \textbf{5.5}$
I	$I = \dfrac{HP \times 746}{E \times E_{ff} \times PF}$	10 HP, 240 V, 90% E_{ff}, 88% PF	I	$I = \dfrac{HP \times 746}{E \times E_{ff} \times PF}$
				$I = \dfrac{10 \text{ HP} \times 746}{240 \text{ V}, 90\% \times 88\%}$
				$I = \textbf{39 A}$

Three-Phase Voltage Values

For 208 V × 1.732, use 360
For 230 V × 1.732, use 398
For 240 V × 1.732, use 416
For 440 V × 1.732, use 762
For 460 V × 1.732, use 797
For 480 V × 1.732, use 831
For 2400 V × 1.732, use 4157
For 4160 V × 1.732, use 7205

Power Formulas — 1ϕ, 3ϕ

Phase	To Find	Use Formula	Example Given	Find	Solution
1ϕ	I	$I = \dfrac{VA}{V}$	32,000 VA, 240 V	I	$I = \dfrac{VA}{V}$ $I = \dfrac{32{,}000\ VA}{240\ V}$ $I = \textbf{133 A}$
1ϕ	VA	$VA = I \times V$	100 A, 240 V	A V	$VA = I \times V$ $VA = 100\ A \times 240\ V$ $VA = \textbf{24,000 VA}$
1ϕ	V	$V = \dfrac{VA}{I}$	42,000 VA, 350 A	V	$V = \dfrac{VA}{I}$ $V = \dfrac{42{,}000\ VA}{350\ A}$ $V = \textbf{120 V}$
3ϕ	I	$I = \dfrac{VA}{V \times \sqrt{3}}$	72,000 VA, 208 V	I	$I = \dfrac{VA}{V \times \sqrt{3}}$ $I = \dfrac{72{,}000\ VA}{360\ V}$ $I = \textbf{200 A}$
3ϕ	VA	$VA = I \times V \times \sqrt{3}$	2 A, 240 V	VA	$VA = I \times V \times \sqrt{3}$ $VA = 2 \times 416$ $VA = \textbf{832 VA}$

Sine Waves

Frequency	Period
$f = \dfrac{1}{T}$ where f = frequency (in hertz) 1 = constant T = period (in seconds)	$T = \dfrac{1}{f}$ where T = period (in seconds) 1 = constant f = frequency (in hertz)

Peak-to-Peak Value

$V_{p\text{-}p} = 2 \times V_{max}$
where
2 = constant
$V_{p\text{-}p}$ = peak-to-peak value
V_{max} = peak value

Average Value

$V_{avg} = V_{max} \times .637$
where
V_{avg} = average value (in volts)
V_{max} = peak value (in volts)
.637 = constant

rms Value

$V_{rms} = V_{max} \times .707$
where
V_{rms} = rms value (in volts)
V_{max} = peak value (in volts)
.707 = constant

System Voltage Ranges*

Supply	Service Range		Point of Use Range	
	Satisfactory	Acceptable	Satisfactory	Acceptable
120, 1φ	114 – 126	110 – 127	110 – 126	106 – 127
120/240, 1φ	114/228 – 126/252	110/220 – 127/254	110/220 – 126/252	106/212 – 127/254
120/208, 3φ	114/197 – 126/218	110/191 – 127/220	110/191 – 126/218	106/184 – 127/220
120/240, 3φ	114/228 – 126/252	110/220 – 127/254	110/220 – 126/252	106/212 – 127/254
277/480, 3φ	263/456 – 291/504	254/440 – 293/508	254/440 – 291/504	254/424 – 293/508

* in volts

Basic Circuit Troubleshooting

Measurement/Test	Expected Results	Possible Problem
Line Voltage (rms)	110 V – 127 V	Low voltage = circuit overloaded, undersized conductor, high resistance splice, too long of circuit run
Peak Voltage	1.41 times rms Voltage	Harmonics producing loads on circuit
Voltage Drop	< 5%	More than 5% = circuit overloaded, undersized conductor, high resistance splice, too long of circuit run
Ground-Neutral Voltage	> 0.5 V – < 5 V	0 V = ground to neutral connection > 5 V = circuit overloaded
Ground Impedence	< 1 Ω to protect personnel < 0.25 to protect electronic equipment	Circuit overloaded, undersized resistance splice, too long of circuit run
GFCI Test	< 7 mA and at a time < 200 ms	Defective GFCI , improper wiring
Arc Fault Test	Breaker trips	Defective AFCI breaker, improper wiring

Electrical/Electronic Abbreviations/Acronyms . . .

Abbr/ Acronym	Meaning	Abbr/ Acronym	Meaning
A	Ammeter; Ampere; Anode; Armature	DPST	Double-Pole, Single-Throw
AC	Alternating Current	DS	Drum Switch
AC/DC	Alternating Current; Direct Current	DT	Double-Throw
A/D	Analog to Digital	DVM	Digital Voltmeter
AF	Audio Frequency	E	Voltage
AFC	Automatic Frequency Control	EMF	Electromotive Force
Ag	Silver	F	Fahrenheit; Fast; Field; Forward; Fuse
ALM	Alarm	FET	Field-Effect Transistor
AM	Ammeter; Amplitude Modulation	FF	Flip-Flop
AM/FM	Amplitude Modulation; Frequency Modulation	FLC	Full-Load Current
		FLS	Flow Switch
ARM.	Armature	FLT	Full-Load Torque
Au	Gold	FM	Fequency Modulation
AU	Automatic	FREQ	Frequency
AVC	Automatic Volume Control	FS	Float Switch
AWG	American Wire Gauge	FTS	Foot Switch
BAT.	Battery (electric)	FU	Fuse
BCD	Binary Coded Decimal	FWD	Forward
BJT	Bipolar Junction Transistor	G	Gate; Giga; Green; Conductance
BK	Black	GEN	Generator
BL	Blue	GRD	Ground
BR	Brake Relay; Brown	GY	Gray
C	Celsius; Capacitance; Capacitor	H	Henry; High Side of Transformer; Magnetic Flux
CAP.	Capacitor		
CB	Circuit Breaker; Citizen's Band	HF	High Frequency
CC	Common-Collector; Configuration	HP	Horsepower
CCW	Counterclockwise	Hz	Hertz
CE	Common-Emitter Configuration	I	Current
CEMF	Counter Electromotive Force	IC	Integrated Circuit
CKT	Circuit	INT	Intermediate; Interrupt
CONT	Continuous; Control	NTLK	Interlock
CPS	Cycles Per Second	IOL	Instantaneous Overload
CPU	Central Processing Unit	IR	Infrared
CR	Control Relay	ITB	Inverse Time Breaker
CRM	Control Relay Master	ITCB	Instantaneous Trip Circuit Breaker
CT	Current Transformer	JB	Junction Box
CW	Clockwise	JFET	Junction Field-Effect Transistor
D	Diameter; Diode; Down	K	Kilo; Cathode
D/A	Digital to Analog	L	Line; Load; Coil; Inductance
DB	Dynamic Braking Contactor; Relay	LB-FT	Pounds Per Foot
DC	Direct Current	LB-IN.	Pounds Per Inch
DIO	Diode	LC	Inductance-Capacitance
DISC.	Disconnect Switch	LCD	Liquid Crystal Display
DMM	Digital Multimeter	LCR	nductance-Capacitance-Resistance
DP	Double-Pole	LED	Light Emitting Diode
DPDT	Double-Pole, Double-Throw	LRC	Locked Rotor Current

. . . Electrical/Electronic Abbreviations/Acronyms

Abbr/ Acronym	Meaning	Abbr/ Acronym	Meaning
LS	Limit Switch	REV	Reverse
LT	Lamp	RF	Radio Frequency
M	Motor; Motor Starter; Motor Starter Contacts	RH	Rheostat
		rms	Root Mean Square
MAX.	Maximum	ROM	Read-Only Memory
MB	Magnetic Brake	rpm	Revolutions Per Minute
MCS	Motor Circuit Switch	RPS	Revolutions Per Second
MEM	Memory	S	Series; Slow; South; Switch
MED	Medium	SCR	Silicon Controlled Rectifier
MIN	Minimum	SEC	Secondary
MN	Manual	SF	Service Factor
MOS	Metal-Oxide Semiconductor	1 PH; 1φ	Single-Phase
MOSFET	Metal-Oxide Semiconductor Field-Effect Transistor	SOC	Socket
		SOL	Solenoid
MTR	Motor	SP	Single-Pole
N; NEG	North; Negative	SPDT	Single-Pole, Double-Throw
NC	Normally Closed	SPST	Single-Pole, Single-Throw
NEUT	Neutral	SS	Selector Switch
NO	Normally Open	SSW	Safety Switch
NPN	Negative-Positive-Negative	SW	Switch
NTDF	Nontime-Delay Fuse	T	Tera; Terminal; Torque; Transf
O	Orange	TB	Terminal Board
OCPD	Overcurrent Protection Device	3 PH; 3φ	Three-Phase
OHM	Ohmmeter	TD	Time Delay
OL	Overload Relay	TDF	Time Delay Fuse
OZ/IN.	Ounces Per Inch	TEMP	Temperature
P	Peak; Positive; Power; Power Consumed	THS	Thermostat Switch
PB	Pushbutton	TR	Time Delay Relay
PCB	Printed Circuit Board	TTL	Transistor-Transistor Logic
PH	Phase	U	Up
PLS	Plugging Switch	UCL	Unclamp
PNP	Positive-Negative-Positive	UHF	Ultrahigh Frequency
POS	Positive	UJT	Unijunction Transistor
POT.	Potentiometer	UV	Ultraviolet; Undervoltage
P-P	Peak-to-Peak	V	Violet; Volt
PRI	Primary Switch	VA	Volt Amp
PS	Pressure Switch	VAC	Volts Alternating Current
PSI	Pounds Per Square Inch	VDC	Volts Direct Current
PUT	Pull-Up Torque	VHF	Very High Frequency
Q	Transistor	VLF	Very Low Frequency
R	Radius; Red; Resistance; Reverse	VOM	Volt-Ohm-Milliammeter
RAM	Random-Access Memory	W	Watt; White
RC	Resistance-Capacitance	w/	With
RCL	Resistance-Inductance-Capacitance	X	Low Side of Transformer
REC	Rectifier	Y	Yellow
RES	Resistor	Z	Impedance

DMM Applications and Features

Recommended Features	Automotive		HVAC		Electrical Maintenance		Electronic Circuit	
	Basic Testing	System Troubleshooting	Basic Testing	System Troubleshooting	Basic Testing	System Troubleshooting	Basic Testing	System Troubleshooting
Bar Graph Display		X		X	X	X	X	X
Capacitance Measurement Mode		X		X		X	X	X
Continuity Test Mode	X	X	X	X	X	X	X	X
Diode Test Mode	X	X				X	X	X
Duty Cycle Mode		X						X
Frequency Counter Mode				X		X	X	X
MIN/MAX Recording Mode		X		X		X		X
Peak MIN/MAX		X		X	X	X	X	X
Relative Mode	X	X	X	X	X	X	X	X
Touch Hold® Mode	X	X	X	X	X	X	X	X
Temperature Measurement		X	X	X		X		X
LoZ			X	X	X	X	X	
LoOhms	X	X		X		X	X	X
LoVAC			X	X	X	X		

Recognized Testing Laboratories (RTLs) and Standards Organizations*

Underwriters Laboratories, Inc.® (UL)	333 Pfingsten Rd. Northbrook, IL 60062 USA Tel: 847-272-8800 www.ul.com
American National Standards Institute (ANSI)	1 W. 42nd St. New York, NY 10036 USA Tel: 212-642-4900 www.ansi.org
British Standards Institution (BSI)	389 Chiswick High Road London W4 4AL United Kingdom www.bsi.org.uk
CENELEC Comité European de Normalisation Electrotechnique	Rue de Stassart, 35 B - 1050 Brussels Tel: +32 2 519 68 71 www.cenelec.be
Canadian Standards Association (CSA)	Central Office 178 Rexdale Blvd. Etobicoke (Toronto), Ont. M9W 1R3 Tel: 416-747-4000 www.csa.ca
Verband der Elektrotechnik und Informationstechnik (VDE)	Frankfurt am main Germany www.vde.de
Japanese Standards Association (JSA)	1-24, Akasaka 4 Minato-ku Tokyo 107 Japan
International Electrotechnical Commission (IEC)	3, rue de Varembé, PO Box 131 1211 Geneva 20 Switzerland Tel: +41 22 919 02 11 www.iec.ch
The Institute of Electrical and Electronic Engineers, Inc. (IEEE)	345 E. 47th St., New York, NY 10017 Tel: 800-678-4333 www.ieee.org
National Institute of Standards and Technology Calibration Program (NIST)	Bldg. 820, Room 232 Gaithersburg, MD 20899 Tel: 301-975-6478 www.nist.gov
National Electrical Manufacturers Association (NEMA) Standards Publication Office	2101 L St., NW Suite 300 Washington, DC 20037 USA Tel: 202-457-8400 www.nema.org
International Standards Organization (ISO)	1 rue de Varembé Case postale 56 CH-1211 Geneva 20 Switzerland Tel: +41 22 749 01 11 www.iso.ch
OSHA Region 1 Regional Office	JFK Federal Building, Room E340 Boston, MA 02203 Tel: 617-565-9860 www.osha.gov
TÜV Rheinland of North America, Inc.	12 Commerce Rd., Newton, CT 06470 Tel: 203-426-0888 www.us.tuv.com

* partial listing

Conductor Color Coding Combinations

Voltage*	Circuit	Conductor Colors
120	1φ, 2-wire with ground	One black (hot wire), one white (neutral wire), and one green (ground wire)
120/240	1φ, 3-wire with ground	One black (one hot wire), one red (other hot wire), one white (neutral wire), and one green (ground wire)
120/208	3φ, 4-wire wye with ground	One black (phase 1 hot wire), one red (phase 2 hot wire), one blue (phase 3 hot wire), one white (neutral wire), and one green (ground wire)
240	3φ, 3-wire delta with ground	One black (phase 1 hot wire), one red (phase 2 hot wire), one blue (phase 3 hot wire), and one green (ground wire)
120/240	3φ, 4-wire delta with ground	One black (first low phase hot wire), one red (second low phase hot wire), one orange (high phase leg wire), one white (neutral wire), and one green (ground wire)
277/480	3φ, 4-wire wye with ground	One brown (phase 1 hot wire), one orange (phase 2 hot wire), one yellow (phase 3 hot wire), one white (neutral wire), and one green (ground wire)
480	3φ, 3-wire delta with ground	One brown (phase 1 hot wire), one orange (phase 2 hot wire), one yellow (phase 3 hot wire), and one green (ground wire)

*in V

Parallel Circuit Calculations

RESISTANCE

$$R_T = \frac{R_1 \times R_2}{R_1 + R_2}$$

where
R_T = total resistance (in Ω)
R_1 = resistance 1 in (in Ω)
R_2 = resistance 2 in (in Ω)

VOLTAGE

$V_T = V_1 = V_2 = \ldots$
where
V_T = total applied voltage (in V)
V_1 = voltage drop across load 1 in (in V)
V_2 = voltage drop across load 2 in (in V) V)

CURRENT

$I_T = I_1 + I_2 + I_3 + \ldots$
where
I_T = total circuit current (in A)
I_1 = current through load 1 (in A)
I_2 = current through load 2 (in A)
I_3 = current through load 3 (in A)

Series Circuit Calculations

RESISTANCE

$R_T = R_1 + R_2 + R_3 + \ldots$
where
R_T = total resistance (in Ω)
R_1 = resistance 1 in (in Ω)
R_2 = resistance 2 in (in Ω)
R_3 = resistance 3 in (in Ω)

VOLTAGE

$V_T = V_1 + V_2 + V_3 + \ldots$
where
V_T = total applied voltage (in V)
V_1 = voltage drop across load 1 (in V)
V_2 = voltage drop across load 2 (in V)
V_3 = voltage drop across load 3 (in V)

CURRENT

$I_T = I_1 = I_2 = I_3 = \ldots$
where
I_T = total circuit current (in A)
I_1 = current through load 1 (in A)
I_2 = current through load 2 (in A)
I_3 = current through load 3 (in A)

Hazardous Locations

Class	Group	Material
I	A	Acetylene
	B	Hydrogen, butadiene, ethylene oxide, propylene oxide
	C	Carbon monoxide, ether, ethylene, hydrogen sulfide, morpholine, cyclopropane
	D	Gasoline, benzene, butane, propane, alcohol, acetone, ammonia, vinyl chloride
II	E	Metal dusts
	F	Carbon black, coke dust, coal
	G	Grain dust, flour, starch, sugar, plastics
III	No groups	Wood chips, cotton, flax, nylon

Voltage Drop Formulas — 1φ, 3φ

Phase	To Find	Use Formula	Example Given	Find	Example Solution
1φ	VD	$VD = \dfrac{2 \times R \times L \times I}{1000}$	240 V, 40 A, 60 L, 0.764 R	VD	$VD = \dfrac{2 \times R \times L \times I}{1000}$ $VD = \dfrac{2 \times .764 \times 60 \times 40}{1000}$ $VD = \mathbf{3.67\ V}$
3φ	VD	$VD = \dfrac{2 \times R \times L \times I}{1000} \times 0.866$	208 V, 110 A, 75 L, 0.194 R, 0.866 multiplier	VD	$VD = \dfrac{2 \times R \times L \times I}{1000} \times 0.866$ $VD = \dfrac{2 \times 0.194 \times 75 \times 110}{1000} \times 0.866$ $VD = \mathbf{2.77\ V}$

$* \dfrac{\sqrt{3}}{2} = .866$

AC/DC Formulas

To Find	DC	AC 1φ, 115 or 220 V	AC 1φ, 208, 230, or 240 V	AC 3φ – All Voltages
I, HP known	$\dfrac{HP \times 746}{E \times E_{ff}}$	$\dfrac{HP \times 746}{E \times E_{ff} \times PF}$	$\dfrac{HP \times 746}{E \times E_{ff} \times PF}$	$\dfrac{HP \times 746}{1.73 \times E \times E_{ff} \times PF}$
I, kW known	$\dfrac{kW \times 1000}{E}$	$\dfrac{kW \times 1000}{E \times PF}$	$\dfrac{kW \times 1000}{E \times PF}$	$\dfrac{kW \times 1000}{1.73 \times E \times PF}$
I, kVA known		$\dfrac{kVA \times 1000}{E}$	$\dfrac{kVA \times 1000}{E}$	$\dfrac{kVA \times 1000}{1.763 \times E}$
kW	$\dfrac{I \times E}{1000}$	$\dfrac{I \times E \times PF}{1000}$	$\dfrac{I \times E \times PF}{1000}$	$\dfrac{I \times E \times 1.73 \times PF}{1000}$
kVA		$\dfrac{I \times E}{1000}$	$\dfrac{I \times E}{1000}$	$\dfrac{I \times E \times 1.73}{1000}$
HP (output)	$\dfrac{I \times E \times E_{ff}}{746}$	$\dfrac{I \times E \times E_{ff} \times PF}{746}$	$\dfrac{I \times E \times E_{ff} \times PF}{746}$	$\dfrac{I \times E \times 1.73 \times E_{ff} \times PF}{746}$

E_{ff} = efficiency

Voltage Conversions		
To Convert	To	Multiply By
rms	Average	9
rms	Peak	1.414
Average	rms	1
Average	Peak	1.567
Peak	rms	707
Peak	Average	637
Peak	Peak-to-Peak	2

Units of Power				
Power	W	ft lb/s	HP	kW
Watt	1	0.7376	341×10^{-3}	0.001
Foot-pound/sec	1.356	1	$.818 \times 10^{-3}$	1.356×10^{-3}
Horsepower	745.7	550	1	0.7457
Kilowatt	1000	736.6	1.341	1

Standard Sizes of Fuses and Circuit Breakers

NEC® 240.6 (a) lists standard ampere ratings of fuses and fixed-trip
Circuit Breakers as follows:
15, 20, 25, 30, 35, 40, 45,
50, 60, 70, 80, 90, 100, 110,
125, 150, 175, 200, 225,
250, 300, 350, 400, 450,
500, 600, 700, 800,
1000, 1200, 1600,
2000, 2500, 3000, 4000, 5000, 6000

Powers of 10

$1 \times 10^4 = 10,000$	$= 10 \times 10 \times 10 \times 10$	Read ten to the fourth power
$1 \times 10^3 = 1000$	$= 10 \times 10 \times 10$	Read ten to the third power or ten cubed
$1 \times 10^2 = 100$	$= 10 \times 10$	Read ten to the second power or ten squared
$1 \times 10^1 = 10$	$= 10$	Read ten to the first power
$1 \times 10^0 = 1$	$= 1$	Read ten to the zero power
$1 \times 10^{-1} = .1$	$= 1/10$	Read ten to the minus first power
$1 \times 10^{-2} = .01$	$= 1/(10 \times 10)$ or $1/100$	Read ten to the minus second power
$1 \times 10^{-3} = .001$	$= 1/(10 \times 10 \times 10)$ or $1/1000$	Read ten to the minus third power
$1 \times 10^{-4} = .0001$	$= \dfrac{1/(10 \times 10 \times 10 \times 10)}{\text{or } 1/10,000}$	Read ten to the minus fourth power

Units of Energy

Energy	Btu	ft lb	J	kcal	kWh
British thermal unit	1	777.9	1.056	0.252	2.930×10^{-4}
Foot-pound	1.285×10^{-3}	1	1.356	3.240×10^{-4}	3.766×10^{-7}
Joule	9.481×10^{-4}	0.7376	1	2.390×10^{-4}	2.778×10^{-7}
Kilocalorie	3.968	3.086	4.184	1	1.163×10^{-3}
Kilowatt-hour	3.413	2.655×10^6	3.6×10^6	860.2	1

Conductive Leakage Current

$$I_L = \frac{V_A}{R_I}$$

where
I_L = leakage current (in microamperes)
V_A = applied voltage (in volts)
R_I = insulation resistance (in megohms)

Temperature Conversions

Convert °C to °F	Convert °F to °C
$°F = (1.8 \times °C) + 32$	$°C = \frac{(°F - 32)}{1.8}$

Resistor Color Codes

Color	Number 1st	Number 2nd	Multiplier	Tolerance (%)
Black (BK)	0	0	1	0
Brown (BR)	1	1	10	—
Red (R)	2	2	100	—
Orange (O)	3	3	1000	—
Yellow (Y)	4	4	10,000	—
Green (G)	5	5	100,000	—
Blue (BL)	6	6	1,000,000	—
Violet (V)	7	7	10,000,000	—
Gray (GY)	8	8	100,000,000	—
White (W)	9	9	1,000,000,000	—
Gold (Au)	—	—	0.1	5
Silver (Ag)	—	—	0.01	10
None	—	—	0	20

Full-Load Currents — DC Motors

Motor Rating (HP)	Current (A) 120 V	Current (A) 240 V
1/4	3.1	1.6
1/3	4.1	2.0
1/2	5.4	2.7
3/4	7.6	3.8
1	9.5	4.7
1 1/2	13.2	6.6
2	17	8.5
3	25	12.2
5	40	20
7 1/2	48	29
10	76	38

Full-Load Currents — 1φ, AC Motors

Motor Rating (HP)	Current (A) 120 V	Current (A) 240 V
1/6	4.4	2.2
1/4	5.8	2.9
1/3	7.2	3.6
1/2	9.8	4.9
3/4	13.8	6.9
1	16	8
1 1/2	20	10
2	24	12
3	34	17
5	56	28
7 1/2	80	40

Full-Load Currents — 3φ, DC Motors

Motor Rating (HP)	208 V	230 V	460 V	575 V
1/4	1.11	0.96	0.48	0.38
1/3	1.34	1.18	0.59	0.47
1/2	2.2	2.0	1.0	0.8
3/4	3.1	2.8	1.4	1.1
1	4.0	3.6	1.8	1.4
1 1/2	5.7	5.2	2.6	2.1
2	7.5	6.8	3.4	2.7
3	10.6	9.6	4.8	3.9
5	16.7	15.2	7.6	6.1
7 1/2	24.0	22.0	11.0	9.0
10	31.0	28.0	14.0	11.0
15	46.09	42.0	21.0	17.0
20	59	54	27	22
25	75	68	34	27
30	88	80	40	32
40	114	104	52	41
50	143	130	65	52
60	169	154	77	62
75	211	192	96	77
100	273	248	1254	99
125	343	312	156	125
150	396	360	180	144
200	—	480	240	192
250	—	602	301	242
300	—	—	362	288
350	—	—	413	337
400	—	—	477	382
500	—	—	590	472

abbreviation: A letter or combination of letters that represents a word.

accuracy: The largest allowable error that occurs under specific operating conditions.

AC sine wave: A symmetrical waveform that contains 360°.

AC voltage: The most common type of voltage used to produce work.

alternating current (AC): Current that reverses its direction of flow at regular intervals.

alternation: One half of a cycle.

ampere: The number of electrons passing a given point in one second.

anode: The positive lead of a diode.

arc blast: An explosion that occurs when the surrounding air becomes ionized and conductive.

arc blast hood: An eye and face protection device that covers the entire head with plastic and material.

arc flash: An electrical discharge that occurs when electrical current passes through the air separating energized/ hot (ungrounded) conductors and ground (grounded parts).

Autorange mode: A DMM function that automatically selects the range with the best accuracy and resolution for the measurement.

autoranging DMM: A DMM that automatically adjusts to a higher range setting if the range is not high enough to provide the least resolution for the measurement taken.

Average Recording mode: A DMM MIN MAX Recording mode that continually calculates the true average of all readings taken over time.

average voltage value (V_{avg}): The mathematical mean of all instantaneous voltage values in a sine wave.

bar graph: A graph composed of individual segments that functions as an analog pointer.

capacitance (C): The ability of a component or circuit to store energy in the form of an electrical charge.

Capacitance Measurement mode: A DMM mode used to measure capacitance or test a capacitor.

capacitor: An electronic device used to store an electric charge.

carbon monoxide (CO): A colorless and odorless toxic gas that is produced when substances that contain carbon (wood, natural gas, gasoline, fuel oil, etc.) are incompletely burned.

carbon monoxide (CO) probe: A DMM accessory used to measure the amount (level) of CO in the air.

cathode: The negative lead of a diode.

clamp-on ammeter: A meter that measures current in a circuit by measuring the strength of the magnetic field around a single conductor.

code: A regulation or minimum requirement.

cold resistance: The resistance of a component when operating current is not passing through the device.

contact temperature probe: A DMM accessory that measures temperature at a single point by direct contact with the area measured.

continuity: The presence of a complete path for current flow.

conventional current flow: Current flow from positive to negative.

crest factor: The ratio of the peak voltage value to the rms voltage value.

current: The amount of electrons flowing through an electrical circuit.

cycle: One complete wave of alternating voltage or current.

DC voltage: Voltage that flows in one direction only.

decibel (dB): The measure of the strength of one electronic signal compared to another.

digital multimeter (DMM): A test tool used to measure two or more electrical values.

diode: An electronic device that allows current to flow in only one direction.

Diode Test mode: A DMM mode used to test diode function.

direct current (DC): Current that flows in one direction only.

DMM accessory: A separate connecting and measuring device used with a DMM to enhance usefulness and efficiency.

DMM specification: Information detailing the capabilities and limitations of a specific DMM.

duty cycle (duty factor): An alternative frequency measurement that is the ratio of time a load or circuit is ON to the time a load or circuit is OFF, expressed as a percentage.

Duty Cycle mode: An alternative Frequency Counter mode that measures the percentage of time a circuit is ON or OFF during a specified period of time.

electrical shock: A condition that results any time a body becomes part of an electrical circuit.

electric arc: A discharge of electric current across an air gap.

electron current flow: Current flow from negative to positive.

Event Logging (Recording) mode: A DMM mode that records circuit measurement data history over a specific time period.

face shield: Any eye and face protection device that covers the entire face with a plastic shield, and is used for protection from flying objects.

fiber optic meter: A DMM accessory used to test fiber optic cable and fittings.

fiber optics: A technology that uses transmitted light signals through flexible fibers of glass, plastic, or other transparent materials.

flash protection boundary: The distance at which PPE is required to prevent burns when an arc occurs.

forward bias: The condition of a diode when a diode allows current flow.

frequency: The number of cycles per second (cps) in an AC sine wave.

Frequency Counter mode: A DMM mode that measures the frequency of AC signals.

generator: An electrical device that converts mechanical energy into electrical energy by rotating a wire coil in a magnetic field.

ghost (stray) voltage: A voltage reading on a DMM that is not connected to an energized circuit.

goggles: An eye protection device with a flexible frame that is secured on the face with an elastic headband.

grounding: The connection of all exposed non-current-carrying metal parts to the earth.

heat: Thermal energy.

hertz (Hz): The international unit of frequency equal to 1 cycle per second.

high-frequency test probe: A DMM accessory that increases the frequency measurement range of the DMM.

high-voltage test probe: A DMM accessory used to increase the voltage measurement range above the DMM listed range.

hot resistance: The actual (true) resistance of a component when operating current is passing through the device.

impedance: The total opposition of any combination of resistance, inductive reactance, and capacitive reactance offered to the flow of alternating current.

inductance (L): The property of a circuit that causes it to oppose a change in current due to energy stored in a magnetic field.

infrared temperature probe: A non-contact temperature probe used for taking temperature measurements by sensing the infrared energy emitted by a material.

Input Alert™: A constant audible warning emitted by a DMM if test leads are connected in current jacks and a non-current function is selected.

International Electrotechnical Commission (IEC): An organization that develops international safety standards for electrical equipment.

lockout: The process of removing the source of electrical power and installing a lock that prevents the power from being turned ON.

Manual Range mode: A DMM mode that allows the selection of a specific measurement range.

maximum resolution: The smallest value that can be discerned on the lowest range setting.

MIN MAX Recording mode (MIN MAX): A DMM mode that captures and stores the lowest and highest measurements for later display.

National Electrical Manufacturers Association (NEMA): A national organization that assists with information and standards concerning proper selection, ratings, construction, testing, and performance of electrical equipment.

National Fire Protection Association (NFPA): A national organization that provides guidance in assessing the hazards of the products of combustion.

noncontact temperature probe: A DMM accessory used for taking temperature measurements on energized circuits or on moving parts.

non-sinusoidal waveform: A waveform that has a distorted appearance when compared with a pure sine waveform.

Ohm's law: The relationship between voltage (E), current (I), and resistance (R) in a circuit.

Occupational Safety and Health Administration (OSHA): A federal government agency established under the Occupational Safety and Health Act of 1970, which requires all employers to provide a safe environment for their employees.

overcurrent: A condition that exists in an electrical circuit when the normal load current is exceeded.

Peak MIN MAX Recording mode: A DMM mode that captures maximum readings of AC sine waveform peak values.

peak-to-peak voltage value (V_{p-p}): The value measured from the maximum positive alternation to the maximum negative alternation.

peak voltage value (V_{max}): The maximum value of either the positive or negative alternation of a sine wave.

period: The time required to produce one complete cycle of a waveform.

personal protective equipment (PPE): Clothing and/or equipment worn by a technician to reduce the possibility of injury in the work area.

polarity: The positive (+) or negative (−) state of an object.

power formula: The relationship between power (P), voltage (E), and current (I) in an electrical circuit.

pressure/vacuum module: A DMM accessory used for taking pressure or vacuum measurements.

rectifier: A device that converts AC voltage to DC voltage by allowing voltage and current to move in one direction only.

Relative mode (REL): A DMM mode that records a measurement and displays the difference between that reading and subsequent measurements.

resistance (R): The opposition to the flow of electrons in a circuit.

resolution: The degree of precise measurement a DMM is capable of making.

response time: The length of time of a new input value required for the DMM to record the new value.

reverse bias: The condition of a diode when it does not allow current flow and acts as an insulator.

root-mean-square voltage value (V_{rms}): Voltage value of a sine wave that produces the same amount of heat in a pure resistive circuit as DC of the same value.

rubber insulating matting: A personal protective device placed on the floor to provide electricians protection from electrical shock when working on energized electrical circuits.

safety glasses: An eye protection device with special impact-resistant glass or plastic lenses, reinforced frames, and side shields.

single-phase AC voltage: Voltage that contains only one alternating voltage waveform.

sinusoidal waveform: A waveform that is consistent with a pure sine wave.

slope: The waveform edge on which the trigger level is selected.

standard: An accepted reference or practice.

symbol: A graphic element that represents a quantity, unit, or component.

tagout: The process of placing a danger tag on the source of electrical power, which indicates that the equipment may not be operated until the danger tag is removed.

temperature: The measurement of the intensity of heat.

Temperature Measurement mode: A DMM mode, with a temperature accessory, that allows direct temperature measurements to be displayed in degrees Fahrenheit or Celsius.

temperature probe: A DMM accessory that measures the temperature of liquids, gases, surfaces, and pipes.

test clip: A DMM accessory attached to a test lead used for making temporary connections for test measurements.

test lead: A flexible insulated lead that serves as a conductor from the accessory to the DMM.

test probe: A DMM accessory used for making electrical contact for test measurements.

thermocouple: A device that produces electricity when two different metals joined together are heated.

three-phase AC voltage: Voltage that is a combination of three alternating voltage waveforms, each displaced 120° (one-third of a cycle) apart.

transducer: A device used to convert physical parameters, such as temperature, pressure, and weight, into electrical signals.

transient voltage (voltage spike): Temporary, undesirable voltage in an electrical circuit.

trigger level: The fixed level at which the DMM counter is triggered to record frequency.

voltage surge: Higher-than-normal voltage that temporarily exists on one or more power lines.

DMM Procedures		
Test/Measurement	Figure	Page
AC Voltage Measurement	5-7	51
Capacitance Measurement	12-2	115
Capacitor Test with Resistance Mode	12-3	117
Clamp-On Ammeter Current Measurement	8-6	80
Continuity Measurement	7-8	72
DC Voltage Measurement	6-4	59
Diode Test	11-3	110
Diode Test with Resistance Mode	11-4	112
Duty Cycle Measurement	10-8	105
Frequency Measurement with Hz Button	10-3	99
Frequency Measurement with Hz Setting	10-2	99
In-Line Ammeter Safety	8-7	82
In-Line Current Measurement	8-8	84
Noncontact Temperature Probe	13-8	129
Resistance Measurement	7-5	68
Testing DMM Fuses	8-5	79

References in italic refer to figures.

DMM Operation Competency Skills . . .

1 SAFETY

_____ 1-1 Identify maximum rated voltage and current listed on DMM terminal jacks and test leads

2 DMM ABBREVIATIONS, SYMBOLS, AND TERMINOLOGY

_____ 2-1 Identify abbreviations on DMM used

_____ 2-2 Identify symbols on DMM used

3 DMM DISPLAYS

_____ 3-1 Read measurement on display with metric prefixes

_____ 3-2 Read bar graph displays

4 DMM ADVANCED FEATURES

_____ 4-1 Use Manual range mode and Autorange mode with measurement obtained

_____ 4-2 Use MIN MAX Recording mode with measurement obtained

_____ 4-3 Use Average mode with measurement obtained

_____ 4-4 Use Relative mode with measurement obtained

_____ 4-5 Use diode test mode

_____ 4-6 Use capacitance measurement mode

_____ 4-7 Use frequency counter mode

_____ 4-8 Use duty cycle mode

_____ 4-9 Use event Logging mode

_____ 4-10 Use AC low pass filter mode

_____ 4-11 Use temperature measurement mode

5 MEASURING AC VOLTAGE

_____ 5-1 Set function switch and connect test leads

_____ 5-2 Connect test leads to test circuit/component

_____ 5-3 Read AC voltage measurement displayed

_____ 5-4 Interpret the significance of AC voltage measurement displayed

6 MEASURING DC VOLTAGE

_____ 6-1 Set function switch and connect test leads

_____ 6-2 Connect test leads to test circuit/component

_____ 6-3 Read DC voltage measurement displayed

_____ 6-4 Interpret significance of DC voltage measurement displayed

7 MEASURING RESISTANCE AND CONTINUITY TESTING

_____ 7-1 (Resistance) Set function switch and connect test leads

_____ 7-2 (Resistance) Connect test leads to test circuit/component

_____ 7-3 (Resistance) Read resistance measurement displayed

_____ 7-4 (Resistance) Interpret significance of resistance measurement displayed

_____ 7-5 (Continuity) Set function switch and connect test leads

_____ 7-6 (Continuity) Connect test leads to test circuit/component

_____ 7-7 (Continuity) Test for continuity

_____ 7-8 (Continuity) Interpret significance of results obtained

8 MEASURING AC AND DC CURRENT

_____ 8-1 (Clamp-On Current Probe) Set function switch and connect test leads

_____ 8-2 (Clamp-On Current Probe) Connect test leads to test circuit/component

_____ 8-3 (Clamp-On Current Probe) Read current measurement displayed

_____ 8-4 (Clamp-On Current Probe) Interpret significance of current measurement displayed

... DMM Operation Competency Skills

_____ 8-5 (In-Line Ammeter) Set function switch and connect test leads
_____ 8-6 (In-Line Ammeter) Connect test leads to test circuit/component
_____ 8-7 (In-Line Ammeter) Read current measurement displayed
_____ 8-8 (In-Line Ammeter) Interpret significance of current measurement displayed

9 OHM'S LAW AND POWER FORMULA
_____ 9-1 Calculate voltage using Ohm's Law
_____ 9-2 Calculate amperage using Ohm's Law
_____ 9-3 Calculate resistance using Ohm's Law
_____ 9-4 Calculate power using Power Formula

10 MEASURING FREQUENCY AND DUTY CYCLE
_____ 10-1 (Frequency) Set function switch and connect test leads
_____ 10-2 (Frequency) Connect test leads to test circuit/component
_____ 10-3 (Frequency) Read frequency measurement displayed
_____ 10-4 (Frequency) Interpret significance of frequency measurement displayed
_____ 10-5 (Duty Cycle) Set function switch and connect test leads
_____ 10-6 (Duty Cycle) Connect test leads to test circuit/component
_____ 10-7 (Duty Cycle) Read duty cycle measurement displayed
_____ 10-8 (Duty Cycle) Interpret significance of duty cycle measurement displayed

11 TESTING DIODES
_____ 11-1 Set function switch and connect test leads
_____ 11-2 Connect test leads to test circuit/component
_____ 11-3 Read measurement displayed
_____ 11-4 Interpret significance of measurement displayed

12 MEASURING CAPACITANCE
_____ 12-1 Set function switch and connect test leads
_____ 12-2 Connect test leads to test circuit/component
_____ 12-3 Read capacitance measurement displayed
_____ 12-4 Interpret significance of capacitance measurement displayed

13 DMM ACCESSORIES
_____ 13-1 Connect and use high-voltage test probe
_____ 13-2 Connect and use temperature probe
_____ 13-3 Connect and use high-frequency test probe
_____ 13-4 Connect and use pressure/vacuum module
_____ 13-5 Connect and use fiber optic meter
_____ 13-6 Connect and use carbon monoxide probe

14 DMM SELECTION
_____ 14-1 Define basic terminology used in DMM specifications

This is to certify that _____ **has received instruction in the skill areas checked. A(n)** _____ **DMM was used in this instruction.**

_____ _____
(Instructor) **(Date)**

Certificate of Completion

This is to certify that _____

has completed _____ hours of instruction in

Basic DMM Principles and Procedures

- DMM Safety Considerations
- Understanding DMM Abbreviations & Symbols
- Reading DMM Displays
- Measuring AC Voltage
- Measuring DC Voltage

- Measuring Continuity & Resistance
- Measuring AC and DC Current (In-Line)
- Measuring AC and DC Current (Clamp-on Current Probe)
- Applying Ohm's Law to Measured Values
- Applying Power Formula to Measured Values

_____
Instructor

_____
Date

FLUKE®

Certificate of Completion

This is to certify that _____

has completed _____ hours of instruction in

Advanced DMM Principles and Procedures

- *Measuring Line Frequency*
- *Testing Diodes*
- *Measuring/Testing Capacitors*
- *Applying MIN/MAX Voltage Measurement Function*
- *Applying MIN/MAX Current Measurement Function*

- *Applying Relative Measurement Function*
- *Using DMM Clamp-On Accessory*
- *Using DMM Temperature Accessory*
- *DMM Selection & Specifications*

Instructor _____

Date _____

FLUKE®